SYSTEMS, SELF-ORGANIZATION AND INFORMATION

Complex system studies are a growing area of central importance to a wide range of disciplines, ranging from physics to politics and beyond. Adopting this interdisciplinary approach, *Systems, Self-Organization and Information* presents and discusses a range of ground-breaking research in complex systems theory.

Building upon foundational concepts, the volume introduces a theory of Self-Organization, providing definitions of concepts including system, structure, organization, functionality, and boundary. Biophysical and cognitive approaches to Self-Organization are also covered, discussing the complex dynamics of living beings and the brain, and self-organized adaptation and learning in computational systems. The convergence of Peircean philosophy with the study of Self-Organization also provides an original pathway of research, which contributes to a dialogue between pragmatism, semeiotics, complexity theory, and self-organizing systems.

As one of the few interdisciplinary works on systems theory, relating Self-Organization and Information Theory, *Systems, Self-Organization and Information* is an invaluable resource for researchers and postgraduate students interested in complex systems theory from related disciplines including philosophy, physics, and engineering.

Alfredo Pereira Jr. is a Professor at Universidade Estadual Paulista Júlio de Mesquita Filho, Brazil.

William Alfred Pickering holds a doctorate in Linguistics, and is currently a freelance translator.

Ricardo Ribeiro Gudwin is an Associate Professor at the University of Campinas, Brazil.

SYSTEMS, SELF-ORGANIZATION AND INFORMATION

An Interdisciplinary Perspective

Edited by Alfredo Pereira Jr., William Alfred Pickering, and Ricardo Ribeiro Gudwin

Routledge
Taylor & Francis Group

LONDON AND NEW YORK

First published 2019
by Routledge
2 Park Square, Milton Park, Abingdon, Oxon OX14 4RN

and by Routledge
711 Third Avenue, New York, NY 10017

Routledge is an imprint of the Taylor & Francis Group, an informa business

British Library Cataloguing-in-Publication Data
A catalogue record for this book is available from the British Library

Library of Congress Cataloging-in-Publication Data
Names: Pereira, Alfredo, Jr., editor. |
Pickering, William A., editor. | Gudwin, Ricardo, 1967– editor.
Title: Systems, self-organization and information: an interdisciplinary perspective / edited by Alfredo Pereira Jr., William A. Pickering, and Ricardo Gudwin.
Description: Abingdon, Oxon; New York, NY: Routledge, 2019.
Identifiers: LCCN 2018030300 | ISBN 9781138609921 (hardback) |
ISBN 9781138609938 (pbk.) | ISBN 9780429465949 (ebook)
Subjects: LCSH: Self-organizing systems. | System theory. |
Information theory.
Classification: LCC Q325 .S944 2019 | DDC 003/.7—dc23
LC record available at https://lccn.loc.gov/2018030300

ISBN: 978-1-138-60992-1 (hbk)
ISBN: 978-1-138-60993-8 (pbk)
ISBN: 978-0-429-46594-9 (ebk)

Typeset in Bembo
by codeMantra

CONTENTS

CONTRIBUTORS

Marcos Antonio Alves, Department of Philosophy, São Paulo State University (UNESP), Marília, Brazil.

Ettore Bresciani Filho, Center for Logic, Epistemology, and the History of Science, University of Campinas (UNICAMP), Brazil.

Mariana Claudia Broens, Department of Philosophy, São Paulo State University (UNESP), Marília, Brazil.

Michel Debrun (1921–1997), Center for Logic, Epistemology, and the History of Science, University of Campinas (UNICAMP), Brazil.

Itala M. Loffredo D'Ottaviano, Center for Logic, Epistemology, and the History of Science, University of Campinas (UNICAMP), Brazil.

Maria Eunice Quilici Gonzalez, Department of Philosophy, São Paulo State University (UNESP), Marília, Brazil.

Ricardo Ribeiro Gudwin, Department of Computer Engineering and Industrial Automation (DCA), School of Electrical and Computer Engineering (FEEC), University of Campinas (UNICAMP), Brazil.

Romeu Cardoso Guimarães, Laboratory of Biodiversity and Molecular Evolution, Department of General Biology, Institute of Biological Sciences, Federal University of Minas Gerais (UFMG), Brazil.

Ivo A. Ibri, Center for Pragmatism Studies, Pontifical Catholic University of São Paulo (PUC-SP), Brazil.

Enidio Ilario, Department of Public Health – Faculty of Medicine, University of Campinas (UNICAMP), Brazil.

Renata Cristina Geromel Meneghetti, Department of Mathematics – Institute of Mathematics and Computer Science, University of São Paulo (USP), Brazil.

João Antonio de Moraes, Department of Philosophy, Faculdade João Paulo II (FAJOPA), Marília, Brazil and Philosophy Teacher at the Instituto Federal de Educação, Ciência e Tecnologia de São Paulo, Votuporanga.

Alfredo Pereira Jr., Department of Education – Institute of Biosciences, São Paulo State University (UNESP), Botucatu, Brazil.

William Alfred Pickering, Center for Logic, Epistemology, and the History of Science, University of Campinas (UNICAMP), Brazil.

Vinicius Romanini, School of Communications and Arts, University of São Paulo (USP), Brazil.

Lauro Frederico Barbosa da Silveira, Department of Philosophy, São Paulo State University (UNESP), Marília, Brazil.

Ricardo Pereira Tassinari, Department of Philosophy, São Paulo State University (UNESP), Marília, Brazil.

Claudia Wanderley, Center for Logic, Epistemology, and the History of Science, University of Campinas (UNICAMP), Brazil.

INTRODUCTION

Alfredo Pereira Jr., William Alfred Pickering, and
Ricardo Ribeiro Gudwin

Complex system studies are a growing area of central importance to a wide range of disciplines. Here, we publish the research from members of the *Interdisciplinary Self-Organization Group* of the Center for Logic, Epistemology, and the History of Science at the State University of Campinas (Campinas, São Paulo State, Brazil). This research group was formed in order to foster the theoretical and applied study of complex systems, and has operated continuously since its foundation. The chapter authors, representing a variety of disciplines, are all members of this group, as well as professors and academic researchers at Brazilian universities. The chapters are structured logically and integrated around the theme of Self-Organization in complex systems, forming a mosaic of different perspectives held together by this central idea. By publishing this collection in English[1], we make these works accessible to an international audience interested in complex systems theory and the related areas of Self-Organization and Information Theory.

The history of this book begins in the 1980s, when French philosopher Dr. Michel Debrun organized a series of seminars to study self-organizing systems. At the same time, two researchers of the Department of Philosophy of the State University of São Paulo (UNESP), located in the city of Marília, São Paulo State, Brazil, went to England to study information theory and cognitive science. Another series of seminars focused on self-organizing systems, organized by Dr. Célio Garcia, was taking place almost 400 miles away in the graduate philosophy program at the Federal University of Minas Gerais (UFMG), located in the city of Belo Horizonte in Minas Gerais State. Members of the three groups joined forces at the end of the 1980s under the leadership of Dr. Debrun. After his death in 1996, Dr. Itala Loffredo D'Ottaviano took his place in the organization of seminars and the coordination of research, tasks that she has continued to perform up to the present time.

The works by Debrun that compose the first two chapters of this book were originally published in Portuguese 1996, but were only recently translated into English. Debrun constructed an original approach to the concept of Self-Organization, using relevant ideas from his predecessors, among them Auguste Cournot, Heinz von Foerster, Hal Ashby, Henri Atlan, Humberto Maturana, Francisco Varela, Ilya Prigogine, and Jean-Pierre Dupuy. A self-organizing system is conceived as an open system that builds its organization and functionality from the patterns of interaction of its components. Self-Organization can coexist with hetero-organization, understood as the case when the organization and functionalities of a system do not derive from the free interaction of the system's components. In consonance with his previous studies on Antonio Gramsci, Debrun argues that a linear hierarchy – as in the case of dictatorial political organizations – is not an instance of Self-Organization, even when the center of power is located inside the system. His concept of Self-Organization requires that the dynamics of a system arise from the free interaction of the components.

The concepts of systems theory used in Debrun's approach to Self-Organization were further developed by D'Ottaviano, a skilled logician, mathematician, and philosopher, with Ettore Bresciani Filho collaborating with his strong background in engineering and administration. They provide apt definitions of concepts such as "system", "structure", "organization", "functionality", and "boundary". The dynamics of self-organizing systems is heavily dependent on the information that is available to and processed by their sub-systems. A cognitive scientist and a philosopher, Maria Eunice Quilici Gonzalez, from the UNESP-Marília group, has joined forces with Alfredo Pereira Jr., a philosopher of science, who participated in the UFMG group to discuss the role of information in Self-Organization. This chapter was originally published in 2008, building on a previous work published in 1996. The authors distinguish informational processes from the properly causal physical processes present in cognitive agents, and attempt to categorize the kinds of information that contribute to self-organizing processes. Extending this framework, we have chosen the remaining chapters from the collections of articles previously published in Portuguese by our group, and have also included other articles by group members that deserve to be presented to an international interdisciplinary community.

The book is divided into four sections. The first section is on foundational concepts, and the second section focuses on biophysical and cognitive approaches to Self-Organization, containing chapters on the complex dynamics of living systems, self-organized adaptation, and learning in computational systems. The third section discusses practical issues of information technology and related ethical questions, all dealt with in the social context of community Self-Organization and technology. The chapters in the final section take a semiotic perspective, investigating the convergence of Peircean philosophy with the study of Self-Organization, an original pathway of research contributing to a dialogue between pragmatism, semeiotics, complexity theory, and self-organizing systems.

The editors are grateful to Itala Maria Loffredo D'Ottaviano, not only for her leadership of the group but also for her help with the book project. We also give our thanks to all chapter authors and other members of the research group for their collaboration in the evolution of the group and contribution to the quality of the results and to FAPESP (Thematic Project 10/52627-9) for funding the research. This is surely an example of a successful collective self-organized process!

The editors hope that this book will not only communicate our group's research to an audience beyond the borders of Brazil, but that it will also demonstrate the wide range of applications of complex systems theory. Above all, we hope that the fruitfulness of the results will inspire readers to further investigations and discoveries in this profound subject of study.

Note

1 Several of the chapters are English translations of previous publications in Portuguese: Chapters 1 and 2: Michel Debrun – A Idéia de Auto-Organização and A Dinâmica da Auto-Organização Primária, in Debrun, M., Gonzalez, M.E.Q., and Pessoa Jr., O. (Orgs.) Auto Organização-Estudos Interdisciplinares, Coleção CLE 18, CLE-UNICAMP, 1996 previously translated by Valéria Venturella; Chapter 3: Ettore Bresciani Filho and Itala Maria Loffredo D'Ottaviano – Conceitos Básicos de Sistêmica, in D'Ottaviano, I.M.L. and Gonzalez, M.E.Q. (Orgs.) Auto-Organização; Estudos Interdisciplinares Coleção CLE 30, CLE-UNICAMP, 2000; Chapter 4: Alfredo Pereira Jr. and Maria Eunice Quilici Gonzalez – O Papel das Relações Informacionais na Auto-Organização Secundária, in Bresciani Filho, E., D'Ottaviano, I.M.L., Gonzalez, M.E.Q. and Souza, G.M. (Orgs.) Auto-Organização; Estudos Interdisciplinares Coleção CLE 52, CLE-UNICAMP, 2008; Chapter 5: Ricardo Pereira Tassinari – Sobre a Realidade-Totalidade como Saber Vivo e a Auto-Organização do Espaço Físico, in Bresciani Filho, E., D'Ottaviano, I.M.L., Gonzalez, M.E.Q., and Souza, G.M. (Orgs.) Auto-Organização; Estudos Interdisciplinares Coleção CLE 52, CLE-UNICAMP, 2008; Chapter 6: Romeu Cardoso Guimarães – Dinâmica Vital, em Bresciani Filho, E., D'Ottaviano, I.M.L., Gonzalez, M.E.Q., Pellegrini, A.M., and Andrade, R.S.C. – Auto-Organização; Estudos Interdisciplinares Coleção CLE 66, CLE-UNICAMP, 2014; Chapter 9: Mariana Cláudia Broens – Auto-Organização e Ação: uma abordagem sistêmica da ação comum, in Bresciani Filho, E., D'Ottaviano, I.M.L., Gonzalez, M.E.Q., and Souza, G.M. (Orgs.) Auto-Organização; Estudos Interdisciplinares Coleção CLE 52, CLE-UNICAMP, 2008; Chapter 11: Renata Cristina Geromel Meneghetti – Uma Compreensão da Economia Solidária à luz da Teoria da Auto-Organização, in Bresciani Filho, E., D'Ottaviano, I.M.L., Gonzalez, M.E.Q., Pellegrini, A.M., and Andrade, R.S.C. – Auto-Organização; Estudos Interdisciplinares Coleção CLE 66, CLE-UNICAMP, 2014; Chapter 13: Vinícius Romanini – Prolegômenos para uma teoria semiótica da Auto-Organização, in Bresciani Filho, E., D'Ottaviano, I.M.L., Gonzalez, M.E.Q., Pellegrini, A.M., and Andrade, R.S.C. – Auto-Organização; Estudos Interdisciplinares Coleção CLE 66, CLE-UNICAMP, 2014; Chapter 15: Lauro Frederico Barbosa da Silveira – Pragmatismo e o Princípio da Continuidade no Cosmos Auto-Organizado, in Souza, G.M., D'Ottaviano, I.M.L. and Gonzalez, M.E.Q. (Orgs.) Auto-Organização: Estudos Interdisciplinares Coleção CLE 38, CLE-UNICAMP, 2004.

PART I
Foundational studies

1

THE IDEA OF SELF-ORGANIZATION

Michel Debrun

The intuition of Self-Organization

The idea of Self-Organization is located at the crossroads between the idea of "organization" and the intuition that we bear about the prefix "self". This term is a linguistic anchor, constantly related to our experiences of the world, particularly to our perception of the interaction – causal, moral, political, or other – between individuals or groups, and to the evaluation that we make of their respective degrees of autonomy and self-affirmation. In these conditions, a definition of Self-Organization, not admissible in Common Sense, in relation to the meaning explicitly or implicitly attributed to "self", would become arbitrary or purposeless. That is what occurs with formulations such as the one proposed by von Foerster (1960), in which Self-Organization is seen as an "increase in the redundancy of the system" or a "decrease in the entropy of a system". Not that those definitions are necessarily wrong. They do not make sense only as long as they cannot be connected to, or rooted in, some intuition, actual or potential, of Common Sense. It could be demonstrated, for example, that the definitions proposed by von Foerster point to an aspect, a condition, or a consequence of Self-Organization, as intuitively defined. This is, therefore, a matter of exploring Common Sense – in the double meaning of unveiling it and using it, systematizing its suggestions or making them more complex, but never overcoming it.

Naturally, that guarantees neither the existence of the phenomena corresponding to these suggestions nor that such phenomena are possible. The concept of Self-Organization might be logically (Fodor, 1980) or logical-mathematically (Ashby, 1962) contradictory. But then, if that were the case, it would be better to renounce the use of the term "Self-Organization" instead of forcefully keeping it to designate, according to certain trends (often found in some scientific areas), phenomena that would be correctly described and explained by other denominations.

Preliminary definition of Self-Organization

Within this "intuitionistic" perspective, we propose an initial, partial, and temporary definition of "Self-Organization": an organization or "form" is self-organized when it produces itself. Considering that each organization is based on discrete elements, it is essential to state that the self-organized form produces itself not in the void but from its very elements. But these elements cannot be of such a nature that their presence mechanically determines the process that has them as a basis. If that were the case, the intuition that we have of "self-production" would be nullified. The conclusion is, then, that the elements constitute just some material and/or foundation, and that what is new or "emerging" in Self-Organization must have its origin at the level of the process itself, not in its initial conditions nor, we must add, in the interchange – material, energetic, informational, symbolic, or other – with the environment.

Definition of "Self-Organization"

Therefore, we arrive at a new definition that clarifies the first one: there is Self-Organization every time the appearance or the restructuring of a form, throughout a certain process, is due to the process itself – and to its intrinsic characteristics – and, only to a lesser degree, to its initial conditions, to the interchange with the environment, or to the casual presence of a supervising instance.

Consequences of the definition

Due to the limited objective of the present chapter, we will postulate that this definition is neither contradictory nor meaningless – although still vague – and that it corresponds to objects that can be either actual or possible. Once this "black box" is accepted, some immediate conclusions can be drawn from the definition:

a   The greater the chasm between the complexity of the final form and the complexity of the sum of the influences (and of casual interactions between these influences) received from the initial conditions and other conditioning circumstances, the greater the degree of the Self-Organization.

b   Self-Organization is always, to some extent, a creation. This does not necessarily mean that it is incompatible with the principle of determinism. If it is compatible (by means of, perhaps, a certain bending or even a redefinition of the idea of determinism), that will force us to speak of "deterministic creation" in the same sense that some speak of "deterministic chaos". But this issue will not be examined here.

c   Considering that Self-Organization is not an absolute issue but a relative one, principles other than that of Self-Organization can intervene beside it, or compete with it, in the organization of a being, of an artifact, or of a situation. In these cases, according to the importance of each principle (planning, for example, or the "Darwinian" combination of blind determinism and chance), Self-Organization can either play the role of "main contractor" or of "sub-contractor".

d   Even though Self-Organization is creation, it continues being a process. It is not an indivisible and almost atemporal act, in contrast to "autopoiesis" (Maturana and Varela, 1980). The latter consists of declarations or definitions that carry out the existence of the object or act to which they refer. For example, if I say, "I promise", I, in fact, promise. Similarly, for Spinoza, the essence of God (or the definition of this essence) – being the essence of "a substance consisting of an infinity of attributes, of which each one expresses an eternal and infinite essence" (Spinoza, 1985, p. 409) – entails the existence of God. Denying it would, in Spinoza's eyes, be contradicting in an absurd way the infinity inscribed in the essence of God. But that is not what happens in a process of Self-Organization. We recognize, however, that the duality proposed here between Self-Organization and autopoiesis does not seem to be consensual for many of the most respected theorists of Self-Organization (see, particularly, Dupuy, 1982).

e   In these conditions, Self-Organization is not a mere consequence of its own beginning. Should that be the case, it would become, precisely, autopoiesis. The beginning would work as a law of construction of what comes next. But the process of Self-Organization only "inherits" this beginning, which it will consider in a very variable way. The beginning is important because it introduces a rupture with the past and with the context, which allows the process as a whole to become independent, in part, from the "remainder of the universe". The beginning also offers orientation or impulse towards a certain direction. In one way or another, it will be integrated into the process, contributing to give it meaning or vigor. It is not known, however, how the previous phases of the process will react to its beginning. The reaction can even be negative.

The meanings of "Self"

But what exactly does this definition mean? The process of Self-Organization is spontaneous; it is "itself".

Three aspects must be pointed out:

a   We have stated that the process is partly autonomous in relation to its starting conditions. That, however, does not entail trivial indeterminism, a capacity to "leap" out of a given situation. On the contrary, in the case of a self-organized process, some of the initial conditions allow the process to fly on its own, to overcome the starting conditions and depart from them. This is different from the evolution of a classical dynamical system, which only expresses or states its initial conditions, even if it results in a "deterministic chaos". For example, let us suppose that among the initial conditions, there are elements of a certain type. On the one hand, they are "actually" and not "analytically" distinct since they are not redundant in relation to one another, that is, because they do not display connections, affinities, and so on, actual or potential, apart from the fact that they are all equally subject to general laws of nature. This causes these elements to "meet" each other, rather than to condition (each other or one and another reciprocally), becoming, therefore, free for new and unheard of connections – which will emerge "*hic et nunc*" – and not only for being actualized or revealed. On the other hand, these elements must be "loose" to a greater or lesser degree. An element will be considered "loose" when, independently of the facts and causalities that preceded its meeting with other "distinct" elements, it "breaks with" or ignores that past. If, for instance, two soccer players feel a certain friendliness for one another, inherited from the past, this feeling will be forgotten or adjourned once the two teams are together on the field. A loose element is an element without memory, disconnected from the context in a general way, and it will only acquire a new memory (that is, participate in the elaboration of a short collective memory, consolidated throughout the game) as a result of its interaction with other distinct and loose elements. Hence, these elements, disconnected from each other and from the remainder of the universe (which is, of course, a limit situation), must invent the formula for their collective Self-Organization, even if there is, occasionally, some kind of determinism ruling the opposition of these two elements or, more plausibly, some kind of determinism being constituted – as determinism – throughout the opposition.

The "conditions of the initial conditions" also contribute to the autonomy of the process. This is a matter of chance in the sense stated by Cournot (1843). These can be casual approaches (or, at the limit, shock, but this possibility is not of interest here) between various elements (such as a "casual series", when, for instance, a series of economic facts "meet" a series of political facts). Or they can be decisions made by individuals, groups, entities, or others (for example, the Brazilian Soccer Confederation decides that a certain game between a certain two teams will happen on a certain day at a certain time). These are chances or decisions that cause really distinct and loose elements – added or not to other kinds of elements – to be gathered at a certain moment in a certain configuration (a stadium or a park, for example), visible or invisible. We will name the set formed by the configuration and the elements included in it an "organization device". The organization device has variable probability of becoming a Self-Organization process (whether or not it is successful). When the device is simply delineated – due to, for instance, the distance between the elements (for example, the initial "great distance" or "reciprocal stanching" between two ideas in someone's mind), we will refer to it as simply a "jumble". Another initial condition that is capable of stimulating or enhancing (or stopping) the autonomy of the process is the frame in which it develops. This frame is formed by institutional dispositions (the definition of legitimate targets, of rules of functioning, of possible sanctions, and so forth), in the case of competitions that can be playful, sportive, economic, political or cultural, and/or actual limits. For example, the process of the Self-Organization of a crowd that tries to escape a building on fire will have a greater probability of being autonomous and creative in relation to these initial conditions if there are multiple possible solutions than if there is just one. In the latter case, the initial conditions tend to rigidly determine the closure, and Self-Organization tends to be replaced by the evolution of a common dynamical system.

In a nutshell, depending fundamentally on itself, that is, being autonomous, is the first condition for an organized process to be "self", to be "itself", to be intelligible to itself. Autonomy with respect to the initial conditions, in turn, is favored by something that, at first sight, seems to be antagonistic to it: the very initial conditions or part of them.

b   But the process is "self" in a second sense: it develops through a task of itself on itself. This second aspect implied by our definition is the most controversial one and is found, directly or indirectly, in the aim of authors such as Ashby (1962) or Fodor (1980). How can we understand this "twist" of the process on itself? Is not this just a play of words? It could be proved, probably, that none of the stages that occur throughout a self-organizing process are rigidly unitary, that such a process always allows for some distance, greater or smaller, depending on the case, between its elements or parts. And that certain parts, at certain moments, can be "more organizing", while other parts are "more organized", with possible interchangeability of roles. This is what makes it possible for the Whole to organize itself, in spite of the conjunctural inequalities.

c   Finally, as it evolves towards the constitution of a form – or towards the restructuring of a given form – the process is "self" or itself in a third sense: the form is not a passive result of the process. The form becomes, through a final organizational adjustment, a resisting *gestalt* (capable of self-reference in certain cases). It has an identity, or it is an identity, which reproduces, in a certain way, in relation to the whole process, the autonomy of this process in relation to its initial conditions.

The engine of Self-Organization

What is the engine of this "self-structure"? What makes it move and causes the Self-Organization process – as a task of itself on itself, whose beginning is a "giver of autonomy" (when it constitutes a real rupture in relation to a previous situation) – to progress and advance towards the constitution or the restructuring of a form?

The main engine of Self-Organization lies in the very interaction between "actually distinct" (and loose) elements, as we suggested earlier, or, we would like to add, in that between "semi-distinct" parts within an organism. In this latter case, the expression "semi-distinct parts" means that the organism is not a holistic being, in which everything fuses to everything else. It means, however, that the parts form an "interiority" or develop an "entangledness", expressed in the fact that each part "knows" about the others, about its own capability to replace them or not, and about its own ability to play one role or another (see Figure 1.1). The problem is to determine how and to what extent this ambiguous situation is compatible with the idea of Self-Organization.

Primary Self-Organization

In the first case of interaction, within a configuration and a possible framework of rules, Self-Organization can be seen as the task of a "macro-agent" of itself on itself, the "process without subject" mentioned by Althusser (1965). This does not mean that this macro-agent has, at least at the starting point, its own objectives, tendencies, and so forth: the only targets – legitimated or not by the frame of interaction – are the participants. It can be, in fact, that some of these elements try to convey a certain orientation to the whole (for example, around a "national project"). But what will actually decide whether or not there is collective Self-Organization is the way in which the proposal is internalized, applied, redefined, diluted, and so forth as the subsequent interactions occur. In any case, the eminent causality of the process is interactive causality, unless, of course, there is the emergence of an agent – individual or collective (for example, a powerful elite) – capable of imposing one course of events or another. But, in that case, it is no longer a matter of Self-Organization but of hetero-organization.

In the same way, when agents associate around a cooperative project (or around the decision to elaborate such a project), this project itself, as a knot uniting all agents' volition, does not guarantee that Self-Organization has occurred or will occur. Volition can be unstable and contradictory. What counts is the sedimentation of "something", which can even be the project itself, or something similar, that would have received the "stamp" of interaction.

As for this first type of Self-Organization, we will state that it is "primary" to point out that it does not depart from an already constituted "form" (a being, a system, and so on) but that, on the contrary, there is the "sedimentation" of a form.

Secondary Self-Organization

When it comes to what occurs in an organism, we will say that secondary Self-Organization takes place when this organism can go, from its own operations, performed on itself, from a certain complexity level – corporal, intellectual,

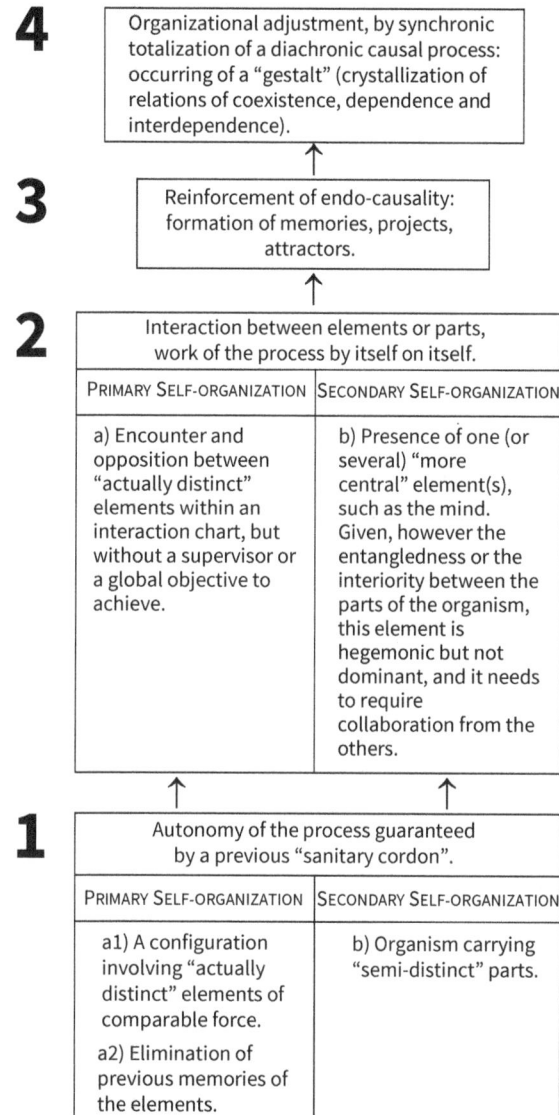

FIGURE 1.1 The process of Self-Organization: general dynamics.

existential, or other – to a superior level. Self-Organization is secondary here since it does begin with simple elements but with an already constituted being or system.

Organisms always display some degree of subjectivity (independently of the way in which we understand subjectivity). In the case of a human organism, we can even perceive the presence of a subject, and we will define this organism as a "subject-form", possessing a "subject-face". A "subject-face" is one that, confronted with an external or internal challenge, "decides", directs, impels, and controls the self-transformation of the organism towards a higher level of complexity. The logical possibility of this passage constitutes the core of the objections raised to the idea of Self-Organization: how can a system, being whatever it is, go to another "ontological" position, counting exclusively or essentially on its own resources? Objectors state that what can be conceived is the progressive maturation of an innate structure that only little by little reveals its potentialities: there is not an increase in complexity but only the manifestation of some complexity given beforehand. We would, in this case, have to abandon the idea of Self-Organization and its creative and constructive character that is defended by many, especially Piaget (1980).

It is not our aim here to deal with this objection, but we can state that it will not be as powerful if, instead of seeing the "subject-face" as omnipotent in relation to the remainder of the organism (as it could be inferred from the previous statements), we emphasize the participation of this "remainder" in the restructuring operations that define secondary Self-Organization. That is, the interaction between parts is present in secondary Self-Organization too. The organism's "subject-face" is one among other parts whose role (and nature) is particularly important but not different from the roles of the other parts. The idea is as follows: due to the combination within the organism of the parts' relative autonomy (particularly, the macro-parts: mind, brain, and "the remainder of the body") and their mutual "entangledness" (each part knows about the others, about their possibility or impossibility of exchanging roles, and so forth), the directive parts can only have – in general and in particular, during the constitutions of new instances of activity – a hegemonic, but not a dominating, role. That is, they do not command from a higher level, or, when they can, it is a matter of an action on peripheral parts of the organism. There is, therefore, a passage from Self-Organization to hetero-organization. The hegemonic role of certain parts means that they "direct", but in order to do that, they need to count on the other parts; otherwise, they cannot do anything. For example, in corporal learning (of a technique or of a sport), the mind has to form images of the body and design such motricity that, "meeting the body", it "fits" (or not) the body's potentialities, causing the body to move (or not) due to a certain resonance effect.

All of this is easy to understand. On the one hand, if there were, among the parts of an organism (mental, cerebral, and others) some complete exteriority, we would be back to primary Self-Organization, unless an element, or a group of elements, emerged as stronger than the others and had the capacity to organize them from a higher level. But then this situation would characterize hetero-organization. On the other hand, if there were an almost-fusion among the parts – as stated by holistic theories such as that developed by Goldstein (1951) – there would not be any possible organizing action: action is always "from" something, "on", or "with" something else; it always implies the existence of real plurality. For Self-Organization to happen "within" the organism, then, there must be among its parts not either radical exteriority or fusion but an intermediate situation, which we call "interiority" or entangledness in the sense presented by Merleau-Ponty (1945). This implies ambiguous and imprecise frontiers, and excludes parts that would be absolute "agents" or "subjects" facing other parts that would be absolute "patients" or "objects". That is, even if there are hierarchies – particularly, the hierarchy mind-body (or "face-subject") – the relations always occur on a basis such as "A in relation to B is more acting than acted upon, and vice-versa". This means that such relation is not of domination but of influence, and it presupposes the participation of the subordinate element. This relation is of "hegemonic" nature, to use the expression coined by Gramsci (1975), although the author limits its use to social, political, and cultural fields.

Naturally, the superior part of the relation (the mind, for instance) may want to "push the other one around", that is, to "command" the other parts (the body) when the latter appears to be hesitating. But then the command will not be obeyed due to the initial relationship between mind and body; to their entangledness; and to the fact that the mind cannot hover over, circumscribe, or control the body.

A new definition of Self-Organization

From this discussion, we can elaborate a new and richer definition of "Self-Organization", one that considers the specificity that the aspect "organization" brings to the aspect "self". The aspect "organization" resides, on the one hand, in the very existence of elements: every organization holds elements which are actually (and not analytically) distinct or, at least, semi-distinct. On the other hand, "self" or "hetero" constitutes an integration of such elements in a "form". Thus, we can say there is Self-Organization every time that, from the encounter of actually (and not analytically) distinct elements, a certain unsupervised interaction (or one without an omnipotent supervisor) occurs and when that interaction eventually results in the constitution of a "form" or in the restructuring by "complexification" of an already existing form.

To this new definition, we can add two auxiliary ones:

a   Primary Self-Organization occurs when the interaction followed by casual integration happens among totally distinct elements (or, at least, among predominantly distinct elements), in a process without subject, central element, or immanent objective. The possible objectives are located at the level of the elements.

b   Secondary Self-Organization occurs when, in a learning process (corporal, intellectual, existential, or other), the interaction occurs between the parts ("mental parts" and/or "corporal parts") of an organism – and the distinction between parts is thus "semi-real" – under the hegemonic, but not dominative, guidance of this organism's "subject-face" (see Figure 1.2).

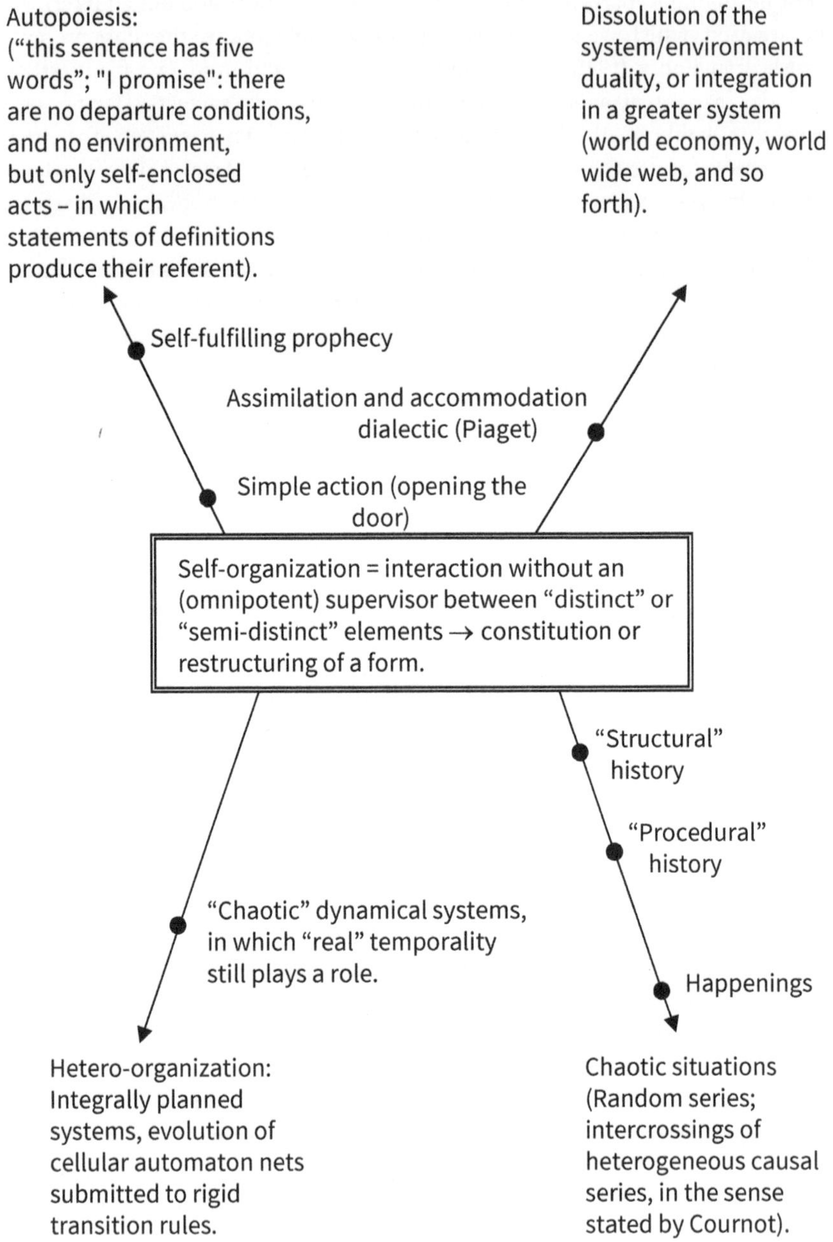

Autopoiesis:
("this sentence has five
words"; "I promise": there
are no departure conditions,
and no environment,
but only self-enclosed
acts – in which
statements of definitions
produce their referent).

Dissolution of the
system/environment
duality, or integration
in a greater system
(world economy, world
wide web, and so
forth).

Self-fulfilling prophecy

Assimilation and accommodation
dialectic (Piaget)

Simple action (opening the
door)

Self-organization = interaction without an
(omnipotent) supervisor between "distinct" or
"semi-distinct" elements → constitution or
restructuring of a form.

"Structural"
history

"Procedural"
history

"Chaotic" dynamical systems,
in which "real" temporality
still plays a role.

Happenings

Hetero-organization:
Integrally planned
systems, evolution of
cellular automaton nets
submitted to rigid
transition rules.

Chaotic situations
(Random series;
intercrossings of
heterogeneous causal
series, in the sense
stated by Cournot).

FIGURE 1.2 Self-Organization and other types of acts, processes, or situations from
the perspective of the degree of control that they exercise upon one
another.

The creativity of Self-Organization

To summarize these statements, we may come to a conception of creativity in Self-Organization, particularly in primary Self-Organization:

a   The simple gathering of actually distinct elements is new; it is almost a creation, as long as, of course, the facts – casual or intentional – that led to this gathering (which we have called "the conditions of the initial conditions") do not exaggerate the scope of their influence within the present. Besides, they must be forgotten; a new "era" is to start. This is what a rupture with relation to the past means. As an example, we may consider that the conversations, decisions, agreements, and so on that caused two boxers to fight lead them to a break with the past, but they cannot influence the fight itself (if, for instance, we learn that one of the fighters agreed, before the fight, not to give his best, then their "encounter" in the ring is not authentic). The fact that constitutes a break with the past cannot become a new past, but it must free the elements that came close to each other (that is, leave them to their own interaction). In short, it all happens as if we had two cascading breaks or as if the rupture happened in two stages. The first break is made by a "breaker", a chance event or decision that we call the "conditions of the initial conditions". The second rupture is a separation, the establishing of a cord between the actually distinct elements (which were gathered) and the breaker itself.

b   Another creativity factor is the nature of the interacting elements. Elements which are "actually distinct" bear, one in relation to the other, a freedom of association that is greater than the one existing between similar elements, once the latter are pre-adjusted by previous affinities, which can be chemical, biological, psychological, and so forth. Then, if distinct elements prevail, the potential for novelty immanent to the Self-Organization will increase to the extent of this prevalence.

c   As the number of distinct elements increases, the degrees of freedom increase as well.

d   The potential for novelty may also be multiplied by the possible heterogeneity of the elements: not all the elements (e.g., soccer players) in primary interaction are homogeneous. The essential factor is that none of them can crush the others. As an example, we may take the interaction between a feeling, a physical nausea, a landscape, a song, or a being inside that landscape, and so forth, as long as there is a supporting element (such as the mind) that acts as a point of meeting, of support, of reciprocal "translation" for the other elements since they belong to different orders of phenomena. "Surrealistic" harmonies may thus emerge, but they should not be seen as the statement or revelation of subjacent pre-established (mystical) harmonies. They emerge and become harmonies "on the spot", that is, at the point where the elements meet and interact. We have here the invention of an organizational adjustment.

e   Within an organism or an already established collectivity, there cannot be, by definition, great heterogeneities since relations of interiority prevail between the elements (which are, because of that, semi-distinct). Even so, an internal creation is possible, and here, it will consist in a relation of interaction/collaboration instead of interaction/competition; each pole will stimulate the other to be "more like itself" because each one needs the other to develop. An instance of this: the relation, discussed by Weber (1930), between capitalism and Protestantism in the two "supports" that Holland or England were in the seventeenth century. In such a case, none of the elements is a mere consequence or expression of the other. Each one possessed relative autonomy within the previous background of a "connection of meaning" (manifestation of reciprocal interiority), which did not exist, for instance, between capitalism and Catholicism. It is understood, in these conditions, that each pole can, at the same time, need the other (since they are different) and assimilate the other's contribution (since they belong to the same universe of meaning). In a nutshell, Protestantism justifies the accumulation and the increase of wealth in contraposition with the "waste" of other eras, whereas the internal and global expansion of capitalism grants the diffusion of some type of Protestantism. We will say that there was creative Self-Organization, the constitution of an original bipolar form. Because, although plausible in the seventeenth-century context, this form was not "inscribed in reality", and in order to emerge, it needed an "organizational adjustment" which constituted a historical invention.

f Finally, the creativity in Self-Organization depends, above all, on the very interaction between elements, both distinct and semi-distinct. We thus return to the beginning of this text. What is decisive is not the intervening mechanisms or the emerging sub-processes in the course of the very process. This will be discussed in detail in another text, dedicated to the dynamics of primary Self-Organization (Debrun, Chapter 2, this volume). We will only remark upon the role played by temporality – or the constitution of a specific temporality for each Self-Organization process. As a complex game develops between real memory (that is, memory not only reconstituted by the observer but also experienced by the system being constituted or redefined) and anticipations based on this memory, the process can at the same time "go forward" and crystallize in a form, progressively creating an attractor and soon obeying it – or, inversely, disobeying it – until a definitive attractor has matured.

Self-Organization and subject

Multiple consequences result from this discussion. Let us concentrate on two of them:

1 The notions or "key words" are not, perhaps, all that would be expected in the intuitive sense that we tend to attribute to "Self-Organization". Let us list these notions:

a Actually distinct elements;
b Meeting, and non-"conditioning", of these elements;

c Interaction of elements, defined in terms of "features of the encounter as such", that is, of the degree of "actual" distinction initially present among the elements;

d Self, with three interconnected meanings, as remarked earlier;

e Primary Self-Organization and secondary Self-Organization;

f Progressive entangledness (or "entangled hierarchy") between a Whole in formation and its parts in primary organization;

g "Entangledness" between the parts of an organism (for example, hierarchic entangledness between mind and body) and reinforcement of these entanglednesses during learning operations in "secondary" Self-Organization.

2 We can notice, in this list of key notions, the absence of the reference to a subject. Could we consider the structuring activity of a subject upon itself, in the sense of a partial or total self-reprogramming, the very paradigm of Self-Organization? Could we interpret what appears in remarks such as "from now on I will redo my life on a completely new basis" as intentions of such kind?

Three remarks are needed here:

a A great part of the studies on Self-Organization (Prigogine and Stengers, 1979; Morin, 1977–1991) have tried to detect Self-Organization mainly at a physical, chemical, or biological level, "on this side of the subject". These studies attempt at rediscovering nature as *physis,* as creating potential, although lacking entities such as the "entelechy" introduced by Driesch (1909) or the "élan vital" introduced by Bergson (1907), which could evoke the notion of a subject or of an "incipient" or "unconscious" subject. The most important thing, for them, is to apprehend nature and some artifacts as being "self" or capable of becoming "self" but not necessarily as a "subject". The subject is "self", almost by definition, but what is "self" is not always "subject".

b We have stated that the subject, or the "subject-face", of certain organisms (such as the human being) intervenes as an initiator, a conductor, and a controller of "secondary" Self-Organization, particularly in learning. But we have also remarked that this "subject-face", or mind, is not omnipotent with relation to the remainder of the organism since it cannot – due to the relation of interiority between them – circumscribe it and hover over it. Indeed, if it could – or when it can – this "subject-face" would – as it does to a local element – convert the remainder of the organism into an object. There would be, then, a break between an organizer Ego and an organizing one, separate from the former. The absolute Self-Organization of a subject by itself, which appears to the eyes of many as the maximization of Self-Organization, would become absolute hetero-organization. Self-Organization, as defined, can only exist in an imperfect way, that is, whenever there is not a "maximization" of the self-organizer subject. Let us add that, even when the causality of the "subject-face" over the remainder of the organism – in "informational" much more than in "energetic" terms – is effective, the entangledness or interiority that exists between them stops the subject pole from understanding in a transparent way the mechanism of its self-organizing operations. This mechanism is obscure to itself, being "stuck" in itself. It does not know, in general, how it moves an arm or how it can perform a sequence of gestures or attitudes. For that reason, many doubt that there can be real interaction of mind and body, and propose that, rather, there must be a parallel relation between them. The recurring return to this thesis does not mean that it is true but that the relation between mind and body is recognized as mysterious and baffling due to their extreme proximity (which does not constitute fusion) and the indefinite frontiers between them.

c   Finally, in "primary" Self-Organization, we often have a multiplicity of competing or collaborating subjects. The competition, "brutal" or based on rules, can be of different types, such as economic, political, sportive, and so forth. The collaboration can be organized around the search for or the implementation of a common project. In any case, the subject – with its calculations, recollections, and anticipations – is everywhere. The problem is that, in many ways, each subject constitutes an obstacle or a limitation for all the others, even when they are in collaboration. The result of this opposition – even when the collective Self-Organization results in a "good form" (not forcefully imposed by one or some of the participants and capable of resistance to possible deviations) – is never of such nature that the individual subject (any of them, including the possible winner) can fully identify with or recognize itself in it. The result does not belong to anyone. There are subjects in Self-Organization, but there is no Subject of Self-Organization. In brief, either the subject is absent in Self-Organization, and, at most, there is – at a physical, chemical, or biological level – a "diffuse" subjectivity, as defined by authors such as Ruyer (1958), or the subject is present only as an element (a main element, of course) of "secondary" Self-Organization; as a *primus inter pares*, that is, first among equals; or as one of the multiple little subjects of the collective human "primary" Self-Organization. In reflecting upon Self-Organization, we never find the subject of Western metaphysics, in charge of itself and of the universe, self-generated, self-transparent, inventor of the moral law (or of the denial of the moral law), donor of meaning to the world. In spite of the multiple intercrossings, there is not much compatibility – and even less fusion – between "philosophy of subject" and the theories of Self-Organization.

Figures 1.1 and 1.2 are synoptic charts that attempt to evoke the general dynamics of Self-Organization. The differences between primary and secondary Self-Organization are shown in Figure 1.1, and the situation of Self-Organization in relation to other acts, processes, and situations – from the perspective of the control that they exercise, or not, upon themselves – is shown in Figure 1.2. For example, autopoiesis has greater control upon itself than Self-Organization since (a) it occurs in a transparent way, (b) it encompasses its referent, and (c) it does not continue or inherit previous elements. It constructs itself from top to bottom, which does not happen in Self-Organization. But the latter, inversely, is more self-controlled than hetero-organized processes or processes submitted to rigorous determinism, which are pushed forward due to their initial conditions.

The top right part of Figure 1.1 refers to the indefinitely extensible spaces where there is no longer "within" or "without", in contrast the internal/external differentiation that occurs progressively in a self-organized process. The notion of self-control, then, becomes meaningless, even when it comes to individuals that move within these spaces and that, at the limit, tend to constitute simple parts of a generalized informational determinism which is not very different – if not in its principles, at least in its results – from energetic determinism. Finally, chaotic situations are defined by the absence of direction, differently from the self-organized processes that acquire or reinforce a central axis.

References

Althusser, L. (1965). *Pour Marx*. Paris: F. Maspero.

Ashby, W. R. (1962). Principles of the self-organizing system. In: von Foerster, H., and Zopf, Jr., G. W. (Eds.), *Principles of Self-Organization* (pp. 255–278). Oxford: Pergamon.

Bergson, H. (1907). *Évolution Créatrice*. Paris: F. Alcan.

Cournot, A. A. (1843). *Exposition de la Théorie des Chances e des Probabilités*. Paris: Hachette.

Driesch, H. A. E. (1909). *Philosophie des Organischen*. Leipzig: Engelmann.

Dupuy, J. P. (1982). *Ordres et Désordres*. Paris: Seuil.

Fodor, J. (1980). Fixation of belief and concept acquisition. In: Piattelli-Palmarini, M. (Ed.), *Language and Learning: The Debate between Jean Piaget and Noam Chomsky* (pp. 142–149). Cambridge, MA: Harvard University Press.

Goldstein, K. (1951). *La Structure de l'Organisme*. Paris: Gallimard. (Original German: Der Aufbau des Organismus.)

Gramsci, A. (1975). *Quaderni del Carcere*. 4 vols. Turim: G. Einaudi.

Maturana, H., and Varela, F. (1980). *Autopoiesis and Cognition: The Realization of the Living. (Boston Studies in the Philosophy of Science 42.)* Boston, MA: Reidel.

Merleau-Ponty, M. (1945). *Phénoménologie de la Perception*. Paris: Gallimard.

Morin, E. (1977–91). *La Méthode*. Paris: Seuil. I. *La Nature de la Nature*, 1977. II: *La Vie de la Vie*, 1980. III: *La Connaissance de la Connaissance*, 1986. IV: *Les Idées – leur Habitat, leur Vie, leurs Moeurs, leur Organisation*, 1991.

Piaget, J. (1980). The psychogenesis of knowledge and its epistemological significance. In: Piattelli-Palmarini, M. (Ed.), *Language and Learning: The Debate between Jean Piaget and Noam Chomsky* (pp. 1–23). Cambridge, MA: Harvard University Press.

Prigogine, I., and Stengers, I. (1979). *La Nouvelle Alliance: Metamorphose de la Science*. Paris: Gallimard.

Ruyer, R. (1958). *La Genèse des Formes Vivantes*. Paris: Flammarion.

Spinoza, B. (1985). *Ethics*. In: *The Collected Works of Spinoza*. Vol. I, Edited and Translated by Edwin Curley. Princeton, NJ: Princeton University Press.

von Foerster, H. (1960). On self-organizing systems and their environments. In: Yovits, M. C., and Cameron, S. (Eds.), *Self-Organizing Systems* (pp. 31–50). Oxford: Pergamon. (Reprinted in von Foerster, H. (1984). *Observing Systems*. 2^a ed. Seaside, CA: Inter-systems, pp. 2–22.)

Weber, M. (1930) *The Protestant Ethic and the Spirit of Capitalism*. London: Routledge.

2

THE DYNAMICS OF PRIMARY SELF-ORGANIZATION

Michel Debrun

We can conceive two types of Self-Organization. In one of them, we have, at the starting point, an organism (or an artifact possessing some of the characteristics of an organism) bearing a "subject-aspect" (and, in the case of the human organism, a "subject-form" having a "subject-face") and aiming, consciously or unconsciously, at restructuring itself to face challenges. That is, it tries to go, through learning, from a certain level of complexity (corporal, intellectual, or existential), to a higher one. What characterizes the "self" here is the fact that both the starting point (the "decision") and the applying point (a certain part of the body, for instance), as well as the mechanisms used and part of the resources, are located within the same organism. The self-organizer subject remains "within itself" during the restructuring operations. It performs a task of itself on itself, which we have defined in previous texts as the nucleus of Self-Organization (Debrun, Chapter 1, this volume).

The initial conditions (the very existence of the organism, the biological, social, and cultural context in which it performs) as well as the interchange with the environment (energetic, material, informational, symbolic, or other) play an important role, but only a supporting one, either through challenges (noise or threatening competition, for instance) or through actual or potential resources offered, or yet through goals suggested as a response to challenges (such as, for example, when Japan copied the West during the Meiji Era in response to the Western imperialistic challenge).

Thus, we move the emphasis, generally placed on the relations established between the self-organized system and its environment (Atlan, 1979), to the "technical" and, we believe, essentially internal operations that constitute the dynamics of this system. This not only constitutes a methodological decision but also raises theoretical problems – which are not going to be approached here – related to the very nature of this type of Self-Organization. For instance, let us consider that the more a system self-organizes, the more it "centers itself on itself", even if it keeps open to the exterior world (and it must do that in order to survive). It can come to constitute a unique or supreme goal, to form a sort of "cyst" in the universe. This conception, which attempts to detect a "logic of closedness" (when other factors do not intervene to hinder or nullify this logic) in Self-Organization, is not necessarily shared by other authors, particularly, by the pioneers of the idea of Self-Organization, such as Atlan.

Second type of Self-Organization: process without a subject

We can also have, at the starting point, a multiplicity of elements that are at the same time "loose" (in relation to each one's past, because this past is "broken" or ignored) and "actually distinct" (that is, lacking logical or casual connections, latent affinities, and so on). In certain circumstances, an interaction – "wild" or within an operational frame – can develop from the encounter of these elements. This interaction may lead to an operational adjustment, to a "form". What is "self" here is the process itself, since it does not derive from previous causalities or affinities or obey an external supervisor, and yet, it results in the constitution of an organization as if it were a planner. This "process without a subject" – in Althusser's (1965) terminology – fulfills in its own way the "task of itself on itself" that characterizes Self-Organization.

Common features between the two types of Self-Organization

In spite of their differences, these two types of process bear several features in common:

a   The already mentioned "task of itself on itself".
b   The fact that this task leads to a growing "centration" of the system on itself, either through the reinforcement and "complexification" of an already existing system or being, or through the constitution of the system itself. The term "centration" here means that the system or being is increasingly "centered in itself", even if this "interest" is completely unconscious.
c   The fact that this "task of itself on itself" entails, in turn, an actual start, that is a start that is not the result of an artificial rupture performed by an observer, but one "inscribed in reality". Otherwise, it would limit itself to extending or reflecting a previous situation, and would not be, by definition, "itself". The rupture can sometimes be minimal when it is a simple "inflection". For instance, in several cases presented by Prigogine and Stengers (1979) in *La Nouvelle Alliance*, a physical, chemical, biological, or other system that is not in a state of equilibrium and that presents slight fluctuations that nullify each other can suddenly present a greater fluctuation that, because it is less likely

than the others to be nullified, will eventually constitute the "knot" of a new self-organized process. It is essential, though, that there is a rupture in relation to the past and also in relation to the context.

d	The two types of Self-Organization presuppose a plurality of elements, and the interaction of these elements is the main engine of Self-Organization (Debrun, Chapter 1, this volume). Whereas the elements of the second mode initially constitute a "dissociated" or "external" plurality in learning, when the organism acts upon itself, the parts (for example, the mind and the body) cannot be completely distinct but only "semi-distinct". Neither of their roles can be rigorously distinct: the mind is "more performing than performed upon" and the body is "more performed upon than performing" – nothing beyond that. Here, we are dealing with an "internal" plurality.

Primary and secondary Self-Organization

When there is an external plurality – which goes from dissociated elements to the constitution of a form – we can say that we are dealing with "primary" Self-Organization (corresponding to the previously explained second type). When, on the other hand, it is a matter of the "self-complexification" of a self-constituted organism (or, more generally, of a system), we are dealing with "secondary" Self-Organization (which corresponds to the first type discussed earlier).

In this chapter, we will be dealing essentially with the dynamics of primary Self-Organization. Some references to the dynamics of secondary Self-Organization will be made, however, to point out either discrepancies or concordances between the two types of Self-Organization. Thus, when we refer simply to "Self-Organization", we will be discussing primary Self-Organization.

The neo-mechanicist perspective

The dynamics of Self-Organization will be addressed within a neo-mechanicist approach, which means that we will assume that:

a	This dynamics, at least at its starting point, does not have a purpose. That is, it is neither stimulated by a central subject (although certain elements may, unsuccessfully, intend to control the process: if that were to happen, Self-Organization would become hetero-organization) nor directed by a previous immanent tendency.

b	Contingent purposes are located at the level of the elements. These elements are, for instance, human individuals that constitute the basis of a great number of playful, sportive, economic, political, and cultural interactions, among others. They are the ones that wish, project, remember, and calculate.

c	These wishes, projections, and so forth, can not only be seen as a meaningful behavior but also as forces that meet, combat and ally, and whose result determines the evolution of a situation up to the point where an organizational adjustment intervenes.

d   However, this mechanicist perspective is termed as "neo" because:

–   It reserves a place, next to significations and energies (and "exchanges" between significations and energies), for the information that the elements emit, receive, and process throughout the interaction.

–   This is, therefore, a matter of elaborating "tridimensional" models, able to integrate the three connections, and not pure "energetic" models, or even "energetic-informational" ones (such as cybernetic models) or "energetic-significational" ones (such as psychoanalytical models). These models must grant, particularly, the understanding of the very interaction that generates an attractor (or a sequence of attractors) able to lead it to organizational adjustment.

–   It is "reductionistic" at the starting point of the explanations (that is, it goes from the elements to the constitution of the "form"), but it becomes progressively "holistic" halfway through the explanation, considering the influence that the increasing consolidation of the form has on the behavior of the elements.

Characteristics of the Self-Organization process

Self-Organization, as it has been defined earlier and in Chapter 1 of this volume, displays some of the characteristics of the organization, in general. In particular, it presupposes actual parts and not only analytical ones. Depending on the case, these parts can be particles, images, sounds, rhythms, individuals, ideas, acts, significations, events, causal series, debris of previous systems, "complete" systems, and so forth. However, in other aspects, Self-Organization displays specific features. For example, in its functional scope:

a   The process of Self-Organization does not always perform functions. What we call primary Self-Organization is "useless", at least in the beginning. It is a matter of "wild" development.

b   When the process, or the partially consolidated system that it becomes, displays functions, these are either exclusively or mainly "self-functions" (that is, performed on behalf of the very process or system) and not "hetero-functions", defined, established, and employed by others, as happens – at least theoretically – in "ordinary" organizations (which are hetero-organizations).

Self-functioning and Self-Organization

As we suggested, Self-Organization is a process that basically develops from itself, without hindering the interchange – material, energetic, informational, symbolic, or other – that it might maintain with the environment. This does not entail the reverse: a process that develops from itself is not always a self-organized process. For example, the passage from a geometrical form to other forms that belong to the same group of transformations cannot be considered a self-organized process, because it does not display interaction of actual elements or parts. It is only an endogenous process that, from its origin, remains always within the same structure. It lacks the creative aspect, which is brought about in the case of Self-Organization by the very interaction of the parts or elements, or by interactions between a partially constituted system and some of its own elements, occasionally reinforced by external elements. The condition is, obviously, this interaction should not be curtailed beforehand by rules and limitations (initial conditions, in a general way) that, instead of stimulating the interaction (as happens in other processes), restrain it excessively. When curtailing happens, we can have self-functioning phenomena, but not, or only very weakly, self-organizing ones.

Let us now introduce the following distinctions:

a   Mechanical automation (clocks, for instance) is excluded from the category of Self-Organization, because it follows a rigid path, without alternative options. Clever automatons that imitate human or animal behavior as conceived by Vaucanson in the seventeenth century (Britannica, 1974), can be very complicated but are not more complex (in terms of "logical type") than the regular clock.

b   One degree above Vaucanson's automatons are situations that maintain or reestablish themselves indefinitely, that is, from their initial force. When an internal or external deviation occurs, it is corrected either immediately or soon. Thus, what we call a "pure energetic circle", without interference of relevant informational components, takes place. We are then back to the starting point. Let us imagine, for example, an initial situation that consists of a great chasm between an elite that concentrates power, wealth, and position, and a population lacking all of that, and the absence of intermediate classes. This situation will tend to remain as such or to be reestablished almost automatically, due to the differential in the initial force that allows the elite to crush or neutralize any manifestation of non-conformism through previous threat, brutal repression, fraud, co-optation of certain elements of the population, and so forth. Such an energetic circle does not display the rigidity of Vaucanson's automatons. More or less unpredictable deviations can emerge. However, an "almost mechanical correction" of these deviations grants the "almost certain" reproduction of the previous situation. In that case, we are not discussing Self-Organization, but the pre-Self-Organization (or a "degree zero" of Self-Organization) of an "almost-being". Or, still, we are discussing some intermediate stage between the evolution of a classic dynamical system and a self-organized process.

c   Cybernetic self-functioning can be seen as "degree 1" of Self-Organization since, besides responding flexibly to the evolution of a situation, it is directed by a norm to be reached or maintained, which might entail the use of alternative means or paths (sometimes not defined beforehand) to reach a certain objective. But the objectives – at least the broadest ones – cannot be redefined by the system itself.

d   An already "crystallized" complex system (the human system, in particular) can redefine its objectives, but not its own being. Here, we have "degree 2" of Self-Organization.

e   There will be Self-Organization as such when a system can "be the very genesis of its own being" (without, however, being able to produce itself).

Definition of primary Self-Organization

The creative feature of the (primary) Self-Organization process can be elucidated in the following definition: there is Self-Organization whenever a process of encounter and interaction of actually distinct elements, without the intervention of a supervisor, or at least an omnipotent supervisor, tends to the constitution of a global form (or a "Whole"), which essentially results from the interaction itself and, only in a lesser degree, from the starting conditions or from interchange with the environment.

Considerations about the definition: the meaning of "actual distinction"

Let us now clarify the already alluded notion of "actual distinction" between elements or parts. This notion relies upon the fact that element A is not "pregnant" with element B, or vice-versa, in logical, causal, or significational terms. That is, neither of the two elements is redundant in relation to the other. Or, alternatively, between them, there cannot be any opposition or structural complementarity, such as that existing between two phonemes. Here lies, as we will often see, the importance of the idea of "encounter": elements that are not conditioned to each other can only have encounters, and these, in turn, always contain some degree of chance. This is the opposite of what occurs, for example, in the logical-mathematical area, where two theorems cannot meet, since they are interconnected beforehand through axioms, derivation rules, and so forth. It is true, though, that "actual distinctness" is not an absolute notion. It can be more or less emphasized depending on the chemical, biological, psychical, or other affinities among the elements, and depending on whether or not these elements belong to the same gravitational, electromagnetic, psychical (both at individual and collective levels), semantical, or other field. Furthermore, as the Self-Organization process evolves, elements can become "interior" to one to another, changing into parts which are not (or no longer, or in a lesser degree) "extra-parts parts". Elements that are all indifferent to one another, that is, lacking affinities or other types of actual or potential bonds, could not, by definition, combine. Their encounter would result in nothing. Some minimal degree of redundancy between certain elements is, therefore, necessary for Self-Organization to happen. Having stated that, we can conceive that the very elements, distinct one in relation to the other, can play the role of "bonds" (we will see that this is the case of the attractor that emerges throughout the process). For this to happen, it is enough that they are surrounded and supported by the others, that is, by associations of related elements, which are thus turned into internal "walls" or "pillars".

Observations on the definition: two limit situations

The more important the interaction becomes, in comparison to the initial conditions of the process, the greater is the degree of Self-Organization, as suggested earlier. Let us now consider two opposite limit situations. On the one hand, the evolution of a close network of cells constituting a "cellular automaton" (Wolfram, 1986): if, in moment t, the state of each cell is determined – according to the Boolean rules established beforehand – by the state of its neighbors in moment t-1, this evolution is rigorously predictable. That is, the interaction of the cells is more apparent than actual, and the Self-Organization within the network is weak (we can refer here to degree zero of Self-Organization), even when there is the emergence of a global and relatively stable form constituted by "islands" of cells that have the same state, separated by zones in which the distribution of the states among the cells seems to be of any sort. On the other hand, let us consider a soccer game, or a conversation without a previous agenda between people who barely know each other. Even if, in both cases, there are rules and explicit or implicit targets coordinating the agents' behavior, and their features and initial states are known beforehand, and so forth, the contingent emergence of a global and persistent form (balance or stagnation in the game or in the conversation or, alternatively, the increasing confirmation of a leadership) depends basically on the very interaction that will be established, on the spot, between the participants. Hence, it is more likely that Self-Organization will either not take place or not be consolidated. But, if it is established, it will constitute a "rich" type of Self-Organization.

Observations on the definition: initial conditions

It is easy to perceive that the "departure" – the degree of autonomy – of the Self-Organization process with relation to its initial conditions depends in great part on these initial conditions as such. So, when these conditions are rigid, as in the case of the evolution of a network of cellular automata, the process tends to be determined in a unilateral way by the initial conditions. Alternatively, in the case of a soccer game, the initial conditions define only a frame (objectives, rules, and so forth), which simultaneously funnels and develops the competition, stimulating the players' creativity.

Phases of the Self-Organization process

We can define three phases – which are, by the way, entangled – in a Self-Organization process when this process does not fail at any point: the beginning, the "interioriza-tion" or "endogenization", and the crystallization of a form or an entity.

First phase: the rupture with a previous situation

The process has a beginning, that is, there is a rupture – greater or smaller, progressive or instantaneous, depending on the case – relative to a previous sit-uation. This rupture can result either from an encounter of various causal series (which forces the development of each series to be interrupted or modified), or from an individual or collective decision (and that, obviously, only happens at the level of human Self-Organization). Hence, the process of Self-Organization:

a   is not the extension or the development of a previous process;
b   is not ruled by an innate code (such as, for instance, the genetic code), at least if this code is understood as a set of rigid rules and impositions;
c   cannot be reduced to the mere maturation or presentation (the Aristotelian passage from "potency" to "act") of an innate structure, differently from what authors such as Chomsky (1980) state about cognitive structures in general and linguistic structures in particular. This is, by the way, the very reason why such authors reject the notion of Self-Organization, which they consider confusing and inapplicable, at least within the cognitive realm;
d   cannot be reduced to the mere application of structures or rules given be-forehand to different contents, even if these contents are always new and unforeseen.

Second phase of the process of Self-Organization: endogenization

The interaction, when successful, is characterized by growing "endogenization". This means the following:

a   As the process advances – until it eventually fails or recedes – the distinction between "within" and "without" is increased.
b   The process is increasingly responsible for its own development, which means that the role of chance, which can be important in the beginning, is progressively "absorbed".

In the case of Self-Organization, in psychological, social, economic, political, or cultural realms, the participants' initial purposes (volitions, intentions, plans, and so forth) will also be absorbed, that is, neutralized, redefined, or subordinated to the global movement of the process. Therefore, the initial situation cannot be seen as the cause of the global process, as "analytically containing" this process. It always plays a fundamental role, since it simultaneously guarantees the singularity of the process and constitutes a "knot", a gradient along which much can happen. Particularly, when it comes to Self-Organization at the human level, the significance of the origins and of the past, in general, is constantly redefined by the present and the projections.

Conditions for endogenization

Endogenization, in turn, requires three conditions, which are stated as follows:

a   The process cannot have an "absolute center" that, from top to bottom, defines its direction and coordinates its development. If that happened, or when that happens, the process would be hetero-organized. There can be, particularly in "secondary" Self-Organization, other types of centers, including a dominating center (such as the brain or the cerebral sub-processes in secondary Self-Organization), but the hierarchy between this center, the other centers, and the process as a whole, can only be an "entangled" hierarchy. The center(s) cannot become exterior in relation to the process. For example, the mental center commands the cerebral one, and the latter commands other parts of the body, but from "within" and within certain limits, without being able to "understand" exactly what is happening. In particular, the mental center cannot be seen, in relation to the body, as "the ghost in the machine", criticized by Ryle (1949), who attributes to Descartes (1998) this radical separation between soul and body. When there are no centers and hierarchies, as in the origins of "primary" Self-Organization, an imprecise equality of forces must reign among the elements that participate in the process (regardless of the exact meaning of the term "force", which changes according to the different areas of Self-Organization). This means that a "winner of the dispute" cannot be designated beforehand. If this were the case, it would again be a matter of hetero-organization.

b   Yet, throughout the process, an attractor must be established due to the very interaction of its elements, and this attractor will make the evolution of the process in a certain direction increasingly likely to happen. This "consolidation" does not always occur, and many self-organized processes fail, that is, the Self-Organization is interrupted halfway. This is because, differently from what happens to dynamical systems in the current sense of the term, which develop towards an attractor given beforehand (a fixed point, a limit cycle, or a chaotic situation). At the beginning of the self-organized process, there is not an attractor awaiting the development or there is only a weak attractor, or even, in other cases (in secondary Self-Organization), there is an "open" and indefinite attractor. What happens is that, in certain cases, depending on conditions and types that we will later attempt to define, an attractor is forged as the process evolves from a "clutter" to a "system".

In other words, stating that the process evolves into a system is the same as asserting that it progressively produces an attractor that represents, at each step and for the further phase, an increasingly pressing requirement. This requirement, depending on the case, can be conceived either in terms of energetic imposition, or (at the human Self-Organization level) in terms of an ideal that is considered desirable, irresistible, ethically ineluctable, and so forth. Let us remark, however, that the attractor is subject to the refluxes that the system can experience while it is not yet consolidated. Even then, from a certain position (which is variable depending on the case, and particularly difficult to secure in the case of human Self-Organization), the attractor, although having been born from the very process, tends to immobilize it.

At the human Self-Organization level, this immobilization can be slowed, re-strained, and sometimes nullified, by external provocations which reopen the access to the world and can, thus, "fabricate negentropy". But, taken by itself, it constitutes an anti-historical factor, that is, it tends to neutralize the "historical coefficient" present in the double fact that primary Self-Organization contains a beginning (which by definition contrasts with previous developments) and that the "endogenization" phase as such is simultaneously creative (in its first sub-phase) and consolidating (only in the second sub-phase).

Given these conditions, we might consider primary Self-Organization (and also secondary Self-Organization, although to a lesser extent) as more semi-historical instead of historical. Being able to integrate historical development as only one aspect among others, Self-Organization cannot be considered totally historical in itself.

c   The constitution of the attractor, in turn, must be based on the presence of an "effective" memory, as defined by Bergson (1970), and not only on a "deduced" memory in the sense of von Foerster (1960) (that is, merely reconstituted by the observer and not "experienced" in the process). The experience of this memory does not solely or necessarily mean that it con-sists of remembrances (or in the potency to evoke them), but that it rests "behind" the present, ready to act as a foundation and as an orientation principle in the immediate future. This memory, therefore, differs from the memory of the processes addressed by classical mechanics, in which the past influences the present but does not remain as such, being, at each moment, absorbed, "engulfed" by the present, and therefore, forgotten as past. On the contrary, in the case of "Bergsonian" memory, the past remains different from the present and yet directly connected to it (that is, it is not necessary to mentally reconstruct it, since it is always "accessible").

Let us recall the example of a soccer game. Suppose that one of the teams has scored two consecutive goals, while the other team has scored none. It is evident that this situation will direct, or "attract", the subsequent steps. The two goals belong now to the collective memories (which are being consolidated throughout the game) of both teams, the referees, the audience, and so on. And this memory is transferred to the future, in the objective form of an indelible result, and in the subjective form of a global expectation, common to all participants, but "expressed" differently according to the categories. This expectation states, generically, the following: it is (much) more probable that team A (which has scored two goals) wins the game than team B (which has not scored any goals). From this, the complex "result + expectation" becomes the outline of an attractor in the face of which the different participants will react according to their momentary dispositions.

In a first eventuality, it is possible that the attractor leads team A to make even greater effort to achieve a brilliant victory, considering that team B is weakening. Simultaneously, the same attractor will lead team B to lose heart. In this case, team A will score more goals, which will reinforce the already outlined attractor. This attractor is similar to a fluctuation, in the sense stated by Prigogine (Prigogine and Stengers, 1979), which, in a system that is far from equilibrium, goes from a small fluctuation to a greater one, and from that to a consolidation, and so on, until the attractor becomes the result, in this case, favorable to team A.

In a second eventuality, the two goals scored by team A may, if projected into the future, lead team A to relax and team B to alarm. In the first moment, just after the goals, the attractor is the same as in the first eventuality, although it leads to different reactions, according to the players' subjective dispositions. In a further moment, however, this attractor may be remodeled, allowing for the outlining of a new attractor. Alternatively, it may collapse from a "gap" in the memory that supported it: there will not be an attractor for the remainder of the game, or at least for part of it, and the Self-Organization process is nullified, allowing for a chaotic situation.

Processes of negative Self-Organization

Naturally, the more closed the initial conditions are, the less we can expect from the memory, the attractor, or the system that is gradually constituted. This closedness intervenes particularly when the initial conditions grant few possibilities of encounters in the future of the process, or, alternatively, when they allow a number of possibilities whose effects tend to nullify or neutralize each other, which admits statistical determinism. This is what happens, for instance, in processes of "negative" Self-Organization. They are based, on the one hand, on the competition of multiple agents for identical targets (such as what happens in low inflation) but not common ones (they cannot or will not associate to combat the low inflation). On the other hand, they are based on the impotence of the agents to act upon one another, to define a winner and a loser (differently from what happens in a game), since there are no teams in which they can find support.

In these terms, each agent, simultaneously, hinders and is hindered by the others in doing something positive. And yet, a Whole can emerge from a solid structure that "bonds" the participants to each other, since they can neither leave the group nor act within it. This Whole is "absent", meaning that it is nowhere, has no visibility, is blurred within the pressure exercised simultaneously by and on all the participants, which blocks all equally. Moreover, as a forming attractor – which, in this case, generally consolidates more quickly than in the case of a soccer game – it drags the participants to a global form of inactivity, such as what happens in "stock market panic", "galloping inflation", and "social or political stagnation or impotence".

We seem now to be returning to a type of classical determinism which, consequently, hinders the possibility of Self-Organization. In reality, this hindering is only tendential, because the possibility of an overturn always remains. This overturn can happen, for instance, due to unpredictable and reciprocal initiatives of some of the participants (after all, according to Sartre, the human being is a "free atom"). It is then possible, as stated in Sartre's *Critique of Dialectical Reason* (1960), that participants go from atomizations, from a "series" of juxtaposed and impotent individuals, to a "group" equipped with some cohesion. This contingency, although not very likely, is enough to assure that the global form, eventually reached through interaction, will be considered as self-organized at least in part, and not merely mechanically produced.

The weakness of the attractor at the beginning of Self-Organization

The absence, the weakness, or the plasticity of the attractor at the beginning of the Self-Organization process derives, in turn, from the diffuse or semi-diffuse character of the elements that do not participate in this process. Either they are still totally dissociated, or they are confronted with a macro-element (a system that is more or less constituted and that will or will not incorporate them), or, yet, they already participate in such a system but find themselves in a spatially or functionally "movable" position. In all these cases, they either meet or miss rather than conditioning each other. This obviously happens in a changeable way, proportional to the degree of their "non-redundancy" with relation to each other. This constitutes, at the limit, a situation of "non-system", from which the absence, the weakness, or the ambiguity of the attractor derives. Considering that this absence makes the Self-Organization process difficult, as suggested earlier, we understand the puzzling character of Self-Organization. It is, however, this very character that contributes to defining Self-Organization as such, and contrasting it, for instance, with the evolution of a common dynamical system, whose components (parameters, variables, particles, or others) are linked beforehand to each other, and where, consequently, there is a "given" attractor.

The distinction between the Whole and its parts

Another aspect of endogenization is that, as the constitution (or the reinforce-ment, or the redefinition) of a form progresses, a division emerges or strengthens. We shall now address the distinction between the Whole and the parts which, in primary Self-Organization remains (since the parts used to be elements) or develops. We shall also discuss, in the case of secondary Self-Organization, the articulation and the distinction that, within a previously existing Whole, is es-tablished between the Whole and the new parts, levels, or functions.

This division, however, cannot become "substantial", that is, it cannot convert the Whole into a substance that hovers over the parts, except in an asymptotic way. In fact, as the Whole and the parts differentiate, the unifying forces of the elements (memory, attractor, and so forth) bring this tendency to a balance, forcing the hierarchy to become a "dirty entangled" one, as mentioned earlier. That is, the emerging hierarchy is of the type in which the poles that respectively become superior and inferior do not show clear frontiers, but ambiguous and movable ones.

Interiority

This "dirty entangled hierarchization" differs from the "clean entangled hierarchi-zation" that exists, for instance, between metalanguage and the object-language in the statement "This statement has five words". The "dirty" hierarchy evokes a notion already discussed in this text: that of "interiority". In a self-organized Whole, or as the Whole organizes itself, the relation between the Whole and its parts, and the relations among the parts themselves, are relations of partial and ambiguous interpenetration – as in a living organism – although probably with smaller complexity, since an organism inherits this property, which has been consolidated throughout the evolution of the species.

This is interiority. So within the living organism, the parts, although distinct (and, because of that, maintaining some possibility of meeting and not only of conditioning each other), "know" each other, regardless of any particular infor-mation that can be transmitted from one to the other. The left hand that pres-sures the right one "knows" that both belong to the same body, that the positions of pressurer and pressure can be inverted at any moment, that both can cooperate to raise an object, and so on (see *Phénoménologie de la Perception*, Merleau-Ponty, 1945). Clearly, the self-organized, or self-organizing, form is less complex than the living form, and so is its "interiority". Even so, there can be interiority, at least an embryonic one.

"Interiority" plays a capital role both in primary and in secondary Self-Organization. In fact:

a   Its presence grants the fulfillment of the first condition of endogeni-zation: the absence of an "absolute" center. In order to be characterized as Self-Organization, a process can only have relative centers, partially "absorbed" in the parts, levels, or functions that they direct or control.

b   Its presence not only hinders the emergence of hierarchies, as we have seen, but also allows them to be "self-hierarchized", or "self-hierarchizations", since they sprout from the Self-Organization process as such and remain consonant with its logic.

c   In the case of learning – learning being the most important manifestation of secondary Self-Organization – we know that an organism intends to transform itself, passing, for instance, from one logical, ontological, or existential level to another. This transformation can only be "self" if the operator and the operated-upon (such as the mind and the arm) maintain a relation of previous interiority that stops the mind from being seen as "pure subject" and the body as "pure object". In order for the Self-Organization process to continue, when it is a matter of linking gestures together, it is essential that the operator does not assume an excessively analytical attitude, which would hinder or break interiority, leading the process to fail or converting Self-Organization into hetero-organization.

The cohesion among the parts

But how can we explain, with regard to interiority, the cohesion that makes it possible and durable, since this cohesion does not exist *ab ovo* in Self-Organization, differently from what occurs in living organisms? Interiority presupposes, on the one hand, the maintenance of an initial relative distinction among the elements that will become parts. Otherwise, there would not be reciprocal interiority but fusion. New distinctions, or even divisions, can emerge throughout the process, as we have just seen. On the other hand, however, we must also understand how the parts may not be of an "extra-parts parts" type (such as "extension", in Descartes (1998) terms), that is, communicating permanently with each other. The previous discussion offers us elements for an answer to this problem, as explained below:

a   We have seen previously that there can be previous affinities among certain elements. If that was all that existed, the actualization of these affinities would be a "gathering" more than an "encounter" (since the related elements display a certain "pre-established harmony"). If a true encounter did not happen, there would not be a true interaction, thus, there would not be true Self-Organization. This is the reason why it is difficult to consider a chemical synthesis as an example of Self-Organization when the synthesis is isolated, when it is not integrated into a broader group of elements and operations. However, an optimum percentage of "related" elements, in relation to "indifferent" ones, is essential in any type of Self-Organization; otherwise, organic "Wholes" would not be achieved, but at best addictive and impermanent Wholes.

b   We stated earlier that the divisions that happen throughout the process – divisions that, theoretically, could shatter the process – are sustained by the unification of elements that takes place previously or concomitantly. That is, the process goes from "One" to multiple without the occurrence of a separation, because it goes simultaneously from multiple to One. The result of this double movement is precisely "dirty entangledness" between the Whole and the parts, and among the parts.

c   We must also take "collective memory" into consideration, even when it is impermanent (as it is during a game) and born among the participants of an interaction when they are not, or at least not all, "inert". Indeed, memory means interpenetrations between the past and the present, with a "tension towards the future". It constitutes, therefore, an integration factor among the participants, since they live and act within this memory.

d   Finally, "negative" Self-Organization as such must be well understood. It resides in the atomization and the impotence of agents in relation to one another, but that does not mean mutual ignorance or stanching, which would mean the same as the Cartesian "extra-parts parts". On the contrary, each agent "knows" about the others, about its incapacity to move them or to associate with them, and about the small possibility of overcoming this situation. Thus, the "bond", which results from the impotence of all in relation to all, as described earlier, develops into interiority, since it is lived by all; this, in turn, "magnifies" this impotence.

Between hetero-organization and the living organism

With relation to the specific problem of "bond", Self-Organization is located, then, in an intermediate situation between hetero-organization and the living organism. In hetero-organization, the "bond" is guaranteed by a centralizing principle, which intervenes from top to bottom and which has, theoretically, the obligation to supervise the cohesion of the edifice at each instant. That entails a tense restless voluntarism, at the risk of seeing everything crumble. In a living organism, inversely, the unification is "given" beforehand, since there is an initial cell whose unity is transmitted, without apparent effort, to all the further unfoldings. Finally, in Self-Organization, the unity does not transcend the parts, nor is it initially given. It emerges – although not always – from the very process in an immanent way, that is, as "adhered" to the process, yet, without being reduced to a purely nominal entity, as are all addictive Wholes.

The ending of the self-organized process

We still need to address an important point: why and how at a certain moment does the process of Self-Organization end?

a   Let us observe that the occurrence of this end means that some type of organization has been reached or is about to be reached. The organization is essentially synchronic: the parts at least coexist, that is, they remain durably juxtaposed, even if they do not do anything for one another. Alternatively, some of them may depend on some others, or all of them may depend on the others and a generalized and stable interdependence takes place.

b   In any case (including the case of hetero-organization, which is fulfilled and controlled by a supervisor), the relations among the parts – up to a point – stop resulting from the pressure performed one on another or from the pressure performed by a supervisor. The process, then, goes from a causal dichotomy to a synchronic arrangement or "adjustment", in which the maintenance of the situation (or its evolution within narrow limits) depends on some type of "agreement" among the parts, even if underlying relations of force continue "guaranteeing" the adjustment.

c   A "hetero-organizational" example might help us better understand this problem. Let us take the case of slave organizations. The fact that the entire population was submitted to slavery through pure violence does not imply that the resulting organization was based solely on pure violence. If that was the case, a regimen of permanent terror should have reigned on the plantation, and slaves' numbers and strength would have made successful escapes and rebellions much more likely than they actually were. In fact, the brutality – but not the terror – that ruled the relations of domination meant that there were, after the slaves' arrival at the plantations, certain phenomena of interlock between them and their masters, which led to some form of "acceptance" by the former. Hence, we can refer here to the notion of "relaxation" or organizational adjustment. This adjustment, depending on the situation, took the form of the slaves' fear (but not necessarily in proportion to the possibility of annihilation by the dominant class, or to the slaves' degree of ignorance about that possibility), their inertia, the affective bonds that they established with their masters, and so on.

d   The problem, then, is the following: in Self-Organization processes – which lack a supervisor (planner, master, and so on) capable of suggesting, facilitating, or stimulating the adjustment – where does the adjustment come from and how does it prevail? It would be pointless to answer that it imposes itself through "the force of things", necessarily, as a mechanical or thermodynamical equilibrium. The difference that we have tried to establish between the evolution of a current dynamical system and a self-organized process relies precisely upon the notion that the latter presents creative aspects. Here, creation seems to depend on the connection, mentioned previously, between memory, anticipation, and formation of attractors.

e   We suggest that the formation of temporary attractors results, at least in human interaction, from anticipation based on memory (for instance, from the short-term memory consolidated during a game): a partial, memorized result is converted into expectation, which outlines an attractor. Such an outline might be confirmed, modified, or abandoned in the following moment, depending on the different ways the agents will react to the attractor. It is possible, however, that at a certain moment the expectation or outline of an attractor – common to all participants, although they have all reacted differently up to that moment – starts inciting identical behavior (resignation, happiness, panic, etc.) in all the participants. This happens because the future, which was projected by a certain expectation, suddenly seems, for all or for most, inevitable (even if, in fact, it is not), irresistible, and desirable, or even irresistible although undesirable. The adjustment will soon intervene (in a soccer game, as we have seen, it can be general conformation or apathy; alternatively, it can be one of the teams winning the game). Obviously, the reasons why a certain future seems to become irresistible to the agents' eyes can appear more or less objective (and sometimes they are not objective at all) to an impartial spectator's eyes. What the agents had was a general impression – rightly or wrongly – that the future, a certain future, was not only probable, but irresistible. This led to the corresponding formation of a definitive attractor. In view of all this, the collective "decision" to conclude the Self-Organization process at such a moment and under such conditions is contingent. Hence it is a creation and not a mechanical result.

f The final acceptance by the participants in a Self-Organization process of a certain type of global situation is comparable to the structure of a "form" as discussed in Gestalt Theory ("*Gestalttheorie*", see Koffka, 1936). Or even better, perhaps the forms discussed by this theory are privileged manifestations of Self-Organization. The emergence of a "*gestalt*" – a psychological one, for example – takes place when its "closedness", through interaction of elementary sensations (amid the support given by an agent's cognitive structures), becomes very high; this probability forces closedness to be anticipated. An instant leap takes place, prepared, but not concluded, by the previous steps. This means that the cognitive agent "runs ahead of itself", trusting (and it can be wrong, since there are permanent examples of it when it comes to identifying an outside object) that the conversion of the highly probable form into the definitive one is certain.

g In the same perspective, it seems plausible to consider the processes that intervene in neural networks as Self-Organization phenomena. It is not only a matter of placing the net itself in orbit or modifying it (through "thickening" or "thinning" of the connections among the net elements, according to "Hebb's rule"), but it is also a matter of its current operations, since these operations seem to involve "anticipations/bets" shared by several brain or "logical" cells in interaction (and sometimes in competition; see Edelman, 1987) just before the perceptive, motor, or other adjustment that will emerge among them.

Subjectivity

If the organizational adjustment is the product of "acceptance/anticipation", and if the adjustment differs from the fixed point of a classical dynamical system, we could perhaps consider that subjectivity and even the philosophical tradition's "subject" is reintroduced everywhere.

In other texts, we have, in fact, limited the subject's role but we cannot refuse "pansubjectivity", even within the "neo-mechanicist" perspective introduced by authors such as Prigogine (see Prigogine and Stengers, 1979) and Morin (1977), who were pioneers in the study of Self-Organization. We share Ruyer's view (as for the atom, for instance) stated in *La Genèse des Formes Vivantes*: We must attribute to the atom the status of an "extension-vision" analogous to a visual sensation, or the status of a duration-melody analogous to an auditory sensation (Ruyer, 1958, p. 65).

All in all, it is enough that, in any field, distanceless relations (although combined with contiguity relations, such as "extra-parts parts" ones) occur among the elements so that they display interiority (in the sense previously defined in this text) between each other, and constitute a being or the outline of a being. This might even be possible in the physical-chemical area.

Absence of Self-Organization in inert elements

We may then state that if "acceptance/anticipation" is essential to organizational adjustment, it entails a minimal amount of subjectivity, and we must refuse the possibility of Self-Organization of inert, inanimate elements, such as matter in the Cartesian conception or a similar one. But that does not exclude the possibility of radically inert things being "gathered" in a self-organized process. Indeed, "indifferent" elements, as mentioned earlier ("inert" and "indifferent" elements are often the same), can be "engaged" as supporting points or "walls", that is, devices capable of granting relative stanching between two regions, and so on.

A new definition for primary Self-Organization

We can now propose a new definition of primary Self-Organization, isomorphic in relation to the one given earlier, but including the preceding discussion: We will define as "primary" Self-Organization every process of integration of actually distinct elements that, instead of tending towards a given attractor, progressively consolidates its own attractor, therefore creating itself as a system.

The corollaries of this new definition are:

a In "primary" Self-Organization, we do not deal with the functioning of a system whose parameters are defined beforehand, but with the genesis of a system from a process, with converting this process into a system.

b   This means that the attractor of the process is the process itself, which tends to "attach to itself", or even to crystallize. The process, as we have seen, is self-referent in a certain way.

c   We have here something similar to the Spinozian tendency of every being to persevere in its being (Spinoza, 1985).

d   This tendency is the initial manifestation of finality. In addition to Spinoza, we can invoke Bergson (1907): it is a matter of "finality without an end", without a target or a plan, at least in the beginning.

Primary and secondary Self-Organization

What has been said up to this point sometimes concerns Self-Organization, in general, but, as has been stated, it particularly concerns primary Self-Organization. For instance, the previously described process of consolidation of an identity or global form is "primary" Self-Organization throughout which an identity develops. "Plenteous" identity (when this plenteousness makes sense) is a finishing line. Let us now compare it to "secondary" Self-Organization, in which a more or less constituted identity (through "primary" organization or any other means) is the starting point for a new process that, although of a new type, maintains several of the aspects discussed earlier.

Primary Self-Organization

In "primary" Self-Organization, identity does not exist in the starting point. That is why it cannot be provoked, disturbed, or stimulated by "noises". Noise, in the sense presented by von Foerster (1960) and Atlan (1979), does not play any role in pure primary Self-Organization, since there is not yet a being or a system to be disturbed by the noise (even though things are, in fact, more -complicated than that, due to the inextricable combination that can take place between "primary" and "secondary" Self-Organization). Nor can we affirm that identity, although non-existing, is "expected" or "desired" from the beginning of the process, or that it will emerge to "perform a function", "fill a void", or any such thing. The elements that are or will be part of this global form can have finalities, intentions, projections, and so forth, when it comes to elements such as human individuals involved in a process of collective Self-Organization. However, at the level of these elements as a group, there is not a previous global finality. And, as we have stated earlier, as the process develops, the characteristic finalities of such elements are neutralized or redefined.

What happens is just that the identity, as it develops, leads to what Spinoza (1985) thought of as the tendency of a substance to remain within its substance, which means that an immanent finality emerges "adhered" to the being, in the sense that, although not pursuing any goal or target, the being "adheres" to its

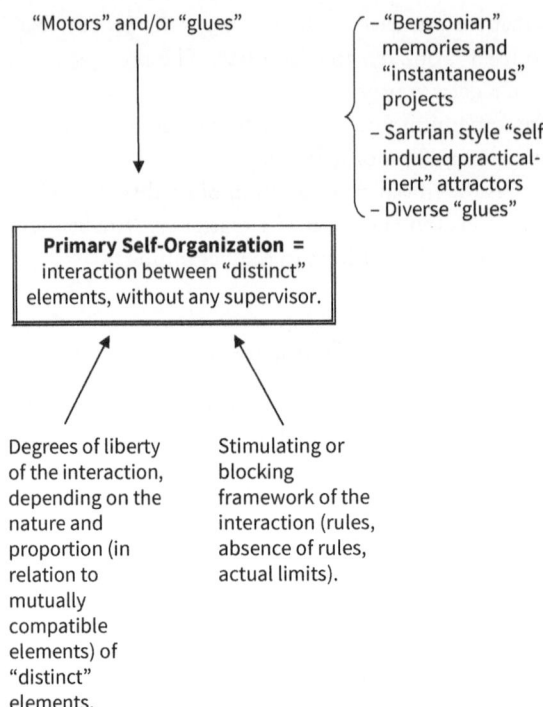

"Motors" and/or "glues"

- "Bergsonian" memories and "instantaneous" projects
- Sartrian style "self-induced practical-inert" attractors
- Diverse "glues"

Primary Self-Organization = interaction between "distinct" elements, without any supervisor.

Degrees of liberty of the interaction, depending on the nature and proportion (in relation to mutually compatible elements) of "distinct" elements.

Stimulating or blocking framework of the interaction (rules, absence of rules, actual limits).

FIGURE 2.1 "Primary" Self-Organization.

own existence. Additionally, and according to the complexity of the system, "orientations", "searches", "intentions", "purposes", "goals", and "functions" can develop in the various parts or operations of an entity (an organism, for example) in relation to the Whole that is being, or has already been, constituted. Thus, in these several cases, we can assume a non-purposive genesis of the finality (Morin, 1977). That is, the finality in its several types appears (sometimes), from within, throughout the organizational process, instead of producing or directing it beforehand (see Figure 2.1).

Secondary Self-Organization

"Secondary" Self-Organization unfolds in a different way. Identity, located in the starting point, is what now "decides" the restructuring of its own being, both within the Whole and within any part, level, or function. The problem, then, is to know how this task of itself on itself is possible not only in a material sense, but also as "self-work".

It is evident, for instance, that trimming the nails or brushing the teeth, although these may, at first sight, seem to be activities of a subject on itself, are, in fact, organizing activities of a subject on exterior objects. Nails and teeth are being treated here as "third-person objects", as if someone else acted upon one from without. That will, indeed, be the fundamental criterion: any activity that an agent performs upon itself, but that could be performed by a third party, cannot be considered a Self-Organization operation. It is a hetero-organizing operation, even if disguised as a self-organizing one. Hence, secondary Self-Organization will only reside in organizational operations in which an agent is the only one that can perform them upon itself. That is the case, for example, of mental or corporal learning or the assimilation of an initially foreign theme or cultural style.

We will discuss the determination of what the requirement of "personalization" of an operation means exactly, and the conditions under which it can be met. Let us just observe here that it must be compatible with another requirement: its fulfillment must not hinder other levels or positions of activity through secondary Self-Organization. In other words, if the personalization of learning operations, transplant of external forms, artistic or literary creation, existential conversion,

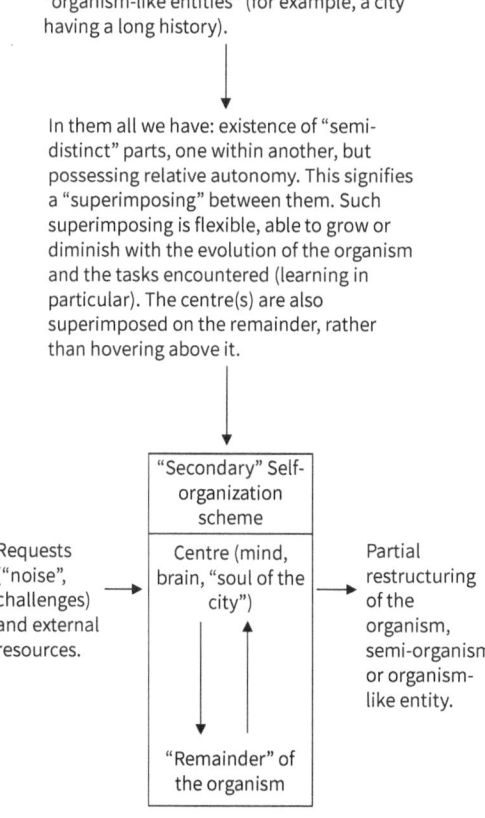

FIGURE 2.2 "Secondary" Self-Organization.

and so on, always entails a certain closedness of the system on itself (a certain self-totalization or "centration" of this system), we must show how that does not hinder – but, rather, facilitates – an "exit from itself" in corporal, intellectual, or existential terms. As an example, we can evoke the "double bind" discussed by Bateson (1971) in *Steps to an Ecology of Mind:* oscillating indefinitely between two experiences – a dive into alcoholism and an ethical/disciplinary struggle against it, for instance – a subject can sometimes "jump up", creating a third position. There is an "exit from itself without exiting itself" that constitutes the paramount of what we call secondary Self-Organization (see Figure 2.2).

Self-Organization and temporality

To conclude, it is important to remark the close relation that Self-Organization dynamics (particularly "primary" Self-Organization) maintains with temporality.

1 Stating that the Self-Organization process does not obey a previously given attractor means that:

    a   Its temporality is effective, and not illusory or semi-illusory. The eventual (and never guaranteed, since it can fail) constitution of the attractor, through lapses and refluxes, demands time. That is, the success of a Self-Organization process is decided *hic et nunc*. It is not virtually enclosed in a previous group of elements, parameters, and variables articulated within a system. The Self-Organization process, when successful, can result in the constitution of a system (the final "form" can be seen as a system or the like of it), but it is not the unfolding or the presentation of a given system.

    b   It could not be greater or smaller than it actually is. If it were greater or smaller, it would be another Self-Organization process or a "non-self-organized" one.

2 Self-Organization is not reproducible, although, in some cases, it can be contingently reproducible. Indeed, in order to be reproducible, it would have to obey a "construction law" (program or algorithm). However, a reproduction dictated by a construction law would not be self-organized, but hetero-organized.

3 Depending on the structure of the temporality – that is, the relation between past, present, and future existing within it – the nature or the degree of Self-Organization will be different. If, for example, the weight of the past is overwhelming – as seems to be the case with biological, vegetable, or animal memory – the possibility of a "secondary" Self-Organization (epigenesis) will be very limited at the level of the individual organism and will demand considerable time for the species. If, inversely, the opening for the future (for the human being) is extreme – under the form of re-Self-Organization projects that attempt to turn the past into *tabula rasa* – these

projects, lacking anchorage, will tend to fail. Or, if they succeed, they will do so because an individual or collective Self-Organization became, in fact, hetero-organization (mutilated in relation to individual personality; dictatorial in relation to the collectivity). The ideal temporal structure for Self-Organization emerges when the past, kept at a certain distance from the present so as not to suffocate it, works as its foundation, helping the agent to project itself towards the future.

4 Self-Organization contains its own beginning, but does not produce it – differently from what happens to autopoiesis and with simple actions (such as opening a door). Both autopoiesis and simple actions are self-centered: in them, the three dimensions of time tend to coincide. In turn, Self-Organization is "lengthened": it has a past that can be perceived *a posteriori* by its participants (when it comes to human beings) without them having generated it in a transparent and intelligible way.

5 As the Self-Organization process tends – when successful – to close on itself due to the consolidation of an attractor, its temporality seems to weaken. The process becomes progressively more visible, until it is converted into a "cyst".

6 That allows us to consider the self-organized process as "semi-historical", since the historicity of the first phase disappears (in a very variable way, according to the situation) in its second phase.

7 We can consider that a "partial and temporary suspension" of determinism takes place at the beginning of the "primary" Self-Organization process. Let us affirm that actually distinct elements, loose in a space-time context, do not, by definition, condition one another. When such elements are predominant in relation to others (actually or potentially connected to one another), their encountering, rather than conditioning, one another allows us to consider that:

a "Much can happen" in the ulterior interaction of these elements (and in their interaction with elements that condition one another). This is the foundation for the notion of "suspension of determinism". This suspension can be seen as a chaotic situation or lapse of time, analogous to what happens when a social or political system decomposes, or when it moves far from equilibrium.

b However, "not all can happen", for already stated reasons: even if they are actually distinct in certain aspects, the elements in interaction belong to one or to many common fields (gravitational, electromagnetic, nuclear, psychological, psychosocial, semantical, "significational", or other). Even if they are in minority, the actually or potentially connected elements can conform to distinct and loose elements, creating supports, "walls", "barriers", and so on, visible or invisible depending on the case. Above all, we know that, in practice, totally distinct and loose elements are passages of an ideal type, and that residues of previous connections (or of previous potential connections) imply residues of deterministic situations prior to the Self-Organization process. That is where the partial character of the suspension of determinism comes from.

c As new connections consolidate throughout the primary Self-Organization process (and, to a lesser degree, the secondary one), determinism is progressively reintroduced, depending on the speed – or lack of it – of the constitution of a definitive attractor. Let us remember that, depending on the type of attractor – a "Bergsonian" one (based on the creative properties of the morphology of temporality, that is, of the different possible combinations between anchorage in the past and tension towards the future), or a "Sartrian" one (constituted by an "inert-practician" of fast crystallization) – this speed will be different.

References

Althusser, L. (1965). *Pour Marx*. Paris: F. Maspero.

Atlan, H. (1979). *Entre le Cristal et la Fumée*. Paris: Seuil.

Bateson, G. (1971). *Steps to an Ecology of Mind*. New York: Chandler.

Bergson, H. (1907). *Évolution Créatrice*. Paris: F. Alcan.

————— (1970). Matière et Mémoire. In: Robinet, A. (Ed.), *Oeuvres*, Edition du Centenaire (pp. 161–382). Paris: PUF, 1970.

Britannica (1974). Vaucanson, Jacques de. In *The New Encyclopaedia Britannica*, 15th ed., Micropaedia (vol. X, p. 369). Chicago, IL: Encyclopaedia Britannica Corporation.

Chomsky, N. (1980). On cognitive structures and their development: a reply to Piaget. In: Piattelli-Palmarini, M. (Ed.), *Language and Learning: The Debate between Jean Piaget and Noam Chomsky* (pp. 751–755). Cambridge, MA: Harvard University Press.

Descartes, R. (1998). Meditations on first philosophy. In: *Discourse on Method and Meditations on First Philosophy* (pp. 45–103). Translated by Donald A. Cress. Indianapolis: Hackett.

Edelman, G. (1987). *Neural Darwinism*. New York: Basic Books.

Koffka, K. (1936). *Principles of Gestalt Psychology*. New York: Harcourt, Brace.

Merleau-Ponty, M. (1945). *Phénoménologie de la Perception*. Paris: Gallimard.

Morin, E. (1977). *La Méthode. I. La Nature de la Nature*. Paris: Seuil.

Prigogine, I., and Stengers, I. (1979). *La Nouvelle Alliance: Metamorphose de la Science*. Paris: Gallimard.

Ruyer, R. (1958). *La Genèse des Formes Vivantes*. Paris: Flammarion.

Ryle, G. (1949). *The Concept of Mind*. London: Hutchinson.

Sartre, J.-P. (1960). *Critique de la Raison Dialectique*. Paris: Gallimard.

Spinoza, B. (1985). Ethics. In: Edited and Translated by Edwin Curley, *The Collected Works of Spinoza* (vol. I) (pp. 399–617). Princeton, NJ: Princeton University Press.

Von Foerster, H. (1960). On self-organizing systems and their environments. In: Yovits, M. C., and Cameron, S. (Eds.), *Self-Organizing Systems* (pp. 31–50). Oxford: Pergamon. (Reprinted *in* von Foerster, H. (1984). *Observing Systems* (2nd ed., pp. 2–22). Seaside, CA: Intersystems.)

Wolfram, S. (1986), *Theory and Applications of Cellular Automata*. Singapore: World Scientific.

3

BASIC CONCEPTS OF SYSTEMICS

Ettore Bresciani Filho and Itala M. Loffredo D'Ottaviano

Introduction

This chapter addresses the fundamental notions, concepts, and definitions of systems science. The designation of this field of knowledge as *systemics* derives from studies developed since the 1950s under the name of *general systems theory*. Starting with the concept of systemic organization, we go on to present the concept of Self-Organization and discuss its interrelation with the concepts of creation and systemic evolution.

A system is here considered as conceived by a subject, who may also ascribe it a finality. But this subject may not currently exist, although at some point it may come to exist; it may, thus, be characterized as a dispositional subject. In this sense, the interpretation of the existence of systems, independently of a subject, is not inconsistent with the existence of systems as deriving from interpretation by a determined subject.

We initially introduce more complex notions, followed by the simpler ones. Many of the relevant terms will be defined in the course of the development of the arguments. Some of the main notions introduced from the perspective of systemics – such as boundary, relation, order, equivalence, and structure – are based on corresponding logical-mathematical concepts, so as to be suited to use in a variety of areas of study (see Shoenfield, 1967). On the other hand, the use of the terms "field" and "force" is an intuitive recourse to ideas taken from the physical sciences.

Beginning in the 1990s, the Center for Logics, Epistemology and the History of Science (CLE) Interdisciplinary Self-Organization Group decided to establish fundamental concepts and nomenclatures necessary for interaction among researchers from a wide variety of fields. In this chapter, however, we have not preoccupied ourselves with giving examples, as our studies on organization and Self-Organization have been elaborated in other works (see D'Ottaviano and Bresciani Filho, 2004a,b; Bresciani Filho et al., 2008).

System

System and elements

A system may be initially defined as a unitary entity of a complex and organized nature, made up of a set of active elements which maintain partial relations between themselves; a system also has characteristics of invariance in time that guarantee its identity. Thus, a *system* is a non-empty set of elements which form a partial structure, with functionality.

In Bresciani Filho and D'Ottaviano (2000), we defined a system as a structure with functionality, that is, as a non-empty set of elements endowed with functionality and with relations. In this chapter, we propose a more general definition which extends this idea, using the concept of partial structure. The concept of partial relation, which allows us to define a partial structure, is presented below.

The non-empty set of elements underlying the system is the system's *universe* or *domain*; note that a system should not be confused with its universe.

A system's *elements* are components, actors, or agents; they are the parts of the system that perform activities (as well as actions, reactions, retroactions, proactions, and transactions), conduct processes and operations, and produce phenomena. The elements are responsible for the system's transformations and conversions, and for events which characterize the system's behavior.

The elements possess characteristics, properties, and predicates which can be expressed by variable or constant parameters. Each parameter can assume values that describe the state of the element. These values are established by the characteristics of the element and by its relations to other elements, and also by the restrictions external to the element or to the system itself.

Elements may be distinguished from one another by the diversity or multiplicity of their individual or relational characteristics and may be classified into three main groups: elements of importation (input), elements of the system's internal transformation processes, and elements of exportation (output).

Functionality is a teleological notion, characterized as a certain informational directioning. It may be related to the system's goals, targets, or ends, and the potential autonomy of the system's components may lead to processes which are not individually, but instead globally, self-organized. The notion of functionality is more basic than that of function, and does not simply correspond to functions developed by the system.

Definition 1. A *system* is a partial structure with functionality, that is, a non-empty set of elements (its *universe*) with partial relations and characterized by a functionality.

The system develops activities (functions, processes, actions), assumes states, and possesses characteristics (properties) of its own.

Subsets of the system's universe can, in their turn, constitute *subsystems*. Such subsets, from the perspective of determined partial relations which characterize the system, also constitute partial substructures of the partial structure underlying the system. Subsystems possess their own functionality, which is an integral part of the functionality of the general system. In particular, a subset containing a single element from the universe may constitute a unitary subsystem of the general system.

The characteristics of the system do not necessarily correspond to the 'sum' (union) of the characteristics of its elements or of its subsystems. Rather, the principle of 'the whole is more, or less, than the sum of its parts', is what defines the property of *synergy* (positive or negative, respectively) of the system.

Among the characteristics of the universe of a system, two are fundamental:

a   The properties and the behavior of each element have effects on the properties and the behavior of the whole, and depend on the properties and the behavior of at least one of the other elements; that is, there are no isolated elements in the system;

b   Each possible subset of elements presents the above characteristic, and thus, the universe set cannot be subdivided into independent subsets.

We emphasize that there may exist elements in a system that are considered neutral. These elements do not develop any activity or function over the course of the existence of the system (or even over part of its existence) and that only belong to the universe of the system. They cannot be understood as isolated elements, however, because their mere presence in some way alters the system's potential for organization or may even alter the organization of the system itself. The characteristics (properties) of the system may be considered emergences (products, resultants). In particular, systemic synergy may be considered the first property which appears in the constitution of a system. Other fundamental properties are globality (constitution of the global unit with its invariance) and the possibility of novelty (in the extreme case, the constitution of the very existence of the system).

System and subject

A system may be considered as an object to be observed, studied, abstracted, conceived, simulated, modeled, or represented by a *subject*, which may not belong to the system's universe. The subject, in the process of representation, seeks knowledge by the comprehension and explanation of the object's existence and properties, and that knowledge can be formalized. Comprehension and signification of this object's (in this case, the system's) existence have a synthetic connotation and fall within the field of the concrete or real, analogical, global, intuitive, and subjective; explanation has an analytical connotation and falls within the field of abstract or imagined, logical, specific, rational, and objective.

The subject, even if it is not an element of the system, establishes an interaction with the object through activities of reflection, speculation, observation, and experimentation. These activities seek qualities of an organization in the object, which characterize its existence, structure, functionality, and possible evolution.

The presence of a subject inevitably implies the presence of a subjective perspective on the system, and not solely an objective one. However, this subjectivity should be seen not in the reductive sense of arbitrary preferences but in the broadened sense of the subject's capacity for interrogation of the object's reality, a capacity inherently constrained by limits of understanding and uncertainty of evaluation. With regard to this point, the importance of interaction between subject and object should be noted.

When the subject is an element of the system's universe, it is as a participant who exerts influence on other elements and is influenced by them; the observer's behavior affects the observed and so on, reciprocally, in a recurrent process. When subject and object are both complex systems, the subject-object interrelation is an interrelation between complex systems. The field of observed (represented) phenomena is defined in the interrelation between subject and object in the domain of form, space, and time.

System and partial relations

Active elements may remain separate or may encounter each other; after encountering, they may separate again, and the process may be repeated continuously or discontinuously. Encounters are essential to the maintenance or establishment of partial relations, direct or indirect.

Partial relations between the elements are manifested in diverse ways: interactions, interrelations, interdependencies, integrations, links, conjunctions, inclusions, implications, identifications, combinations, connections, and in other ways as well.

These relations exert restrictions, make impositions, and establish underlying regularities of the elements' activities in the form of laws and rules, decision hierarchies, regularity control, equilibrium adjustment, and command over changes. For the relations to be interpreted as interactive reciprocal action between elements, it is supposed that in the system there are active elements with autonomy or the possibility of encountering one another.

Encounters may be determined or undetermined. When they occur in a predictable way they characterize necessity; otherwise, when they result from the unpredictable, uncertain, or undetermined, they denote randomness; but even in the undetermined encounters, elements are submitted to certain restrictions that depend on their natures and on partial relations already existing in the system, and this too establishes necessity. It may be observed that the greater the number of restrictive situations, the lesser the degrees of freedom or autonomy of elements constitutive of the system.

From a random distribution of elements there may occur encounters which originate an important systemic characteristic, *organization*. In a situation of non-regularity, there may occur encounters which determine partial relations and create organization. This situation may change again, moving towards disintegration and disorganization, only to return afterwards to the constitution of organization (in this case, reorganization or Self-Organization) in a circular process of recurrent transformation.

It is possible to establish four distinct types of relations: relations of the first type, which are functionally necessary, characterize the system; relations of the second type, referred to as positive synergetic or cooperation relations, are complementary – they add something to the system's behavior; relations of the third type, called negative synergetic or competition relations, indicate conditions of antagonism, discordance, and opposition among elements; relations of the fourth type, called redundant relations, duplicate the existing ones. Partial relations may be of more than one type; they may change over the course of the dynamic existence of the system and may even change their type. The presence of distinct types of relations guarantees the possibility of emergences, which include the possibility of self-organizing processes.

The system's elements constitute a network of partial relations, which in general arrange themselves in treelike and circular relations; the latter may be oriented forwards (proaction) or backwards (retroaction). Partial relations of hierarchy are particular cases of the treelike and order relations.

Circular relations (ring or ribbon ones) are supported on the principle of the recurring cycle, identified as processes of the following types: those in which the effects of a relation between elements are causes of this same relation; those in which the product of a system affects the process of its production; those in which the final state generates or modifies the initial state of the system; or those in which effects retroact on their causes.

Partial relations and order

The general notion of relation used in the definition of system, that of *partial relation*, is an extension of the usual logical-mathematical concept of relation. The concept of partial relation, presented by Mikenberg et al. (1986) as basic to the introduction of the mathematical concept of *pragmatic truth*, later called *quasi-truth* by da Costa, has recently received various applications in logic and philosophy of science (see D'Ottaviano and Hifume, 2007; D'Ottaviano, 2010).

Quasi-truth intends to accommodate the incompleteness inherent in scientific representations, and to access the meaning of theories of truth of pragmatist philosophers like James, Dewey and, particularly, Peirce (see Peirce, 1958). The definition of quasi-truth, not discussed here, was proposed by da Costa as a generalization of Tarski's (1944) formal characterization of truth for partial contexts. Starting with the concept of partial relation, which generalizes the concept of relation, Tarski's concepts of structure and truth are extended by the introduction of the concept of partial structure, which supports our definition of system.

An *n-ary relation* on a given set is any subset whose elements are finite (*n*-uple) sequences of elements of the set. As a particular case, *binary relations* correspond to subsets of the set of ordered pairs of elements of the initial set; if an ordered pair belongs to a binary relation, it is said that the pair satisfies the given relation – the first element of the pair is in relation to the second.

A *structure* is a non-empty set – its *domain* or *universe* – with relations between its elements.

According to da Costa, when a certain domain of knowledge (a universe) is investigated, we submit it to a conceptual scheme, with the goal of systematizing and organizing information about it. This domain is represented by a set of objects D and studied by the analysis of the relations between its elements. Given a relation R, on D, it is usual in scientific contexts that "we do not know" if all objects of D are related to R; we say that our information relative to the domain of knowledge is incomplete or partial. For da Costa, the notions of partial relation and partial structure enable us to formally accommodate such incompleteness and to represent the information about the domain of investigation.

We, therefore, have adopted the concept of partial relation, which seems adequate to a general definition of system and even of Self-Organization.

Definition 2. Let D be a non-empty set. An *n-ary partial relation* R on D is an ordered term $\langle R_1, R_2, R_3 \rangle$, R_1, R_2, *and* R_3 with no two of them having common elements, and whose union is D $(R_1 \cup R_2 \cup R_3 = D)$, such that:

1 R_1 is the set of the n-tuples that we know belong to R;
2 R_2 is the set of the n-tuples that we know do not belong to R;
3 R_3 is the set of the n-tuples for which we do not know if they belong to R or not.

We observe that if R_3 does not have any elements $(R_3 = \varnothing)$, then R is a usual n-ary relation, identified with R_1.

Definition 3. A *partial structure* is a pair $\langle D, R_i \rangle_{i \in I}$. with D a non-empty set and each R_i, $i \in I$, a partial relation on D.

The partial relations which characterize the partial structure and the activity of the system may be of distinct natures. Among them, there may occur relations of order. By definition, a binary *partial relation* on a set is *of order* when it is reflexive, anti-symmetric, and transitive: for reflexivity, every element of the set is in relation to itself, i.e., every pair of identical elements satisfies the relation; for anti-symmetry, if two elements of the set are such that the first is in relation to the second and the second is in relation to the first, then they are identical; for transitivity, if three elements of the system are such that the first is in relation to the second and the second is in relation to the third, then the first is in relation to the third.

We can also consider asymmetric partial relations of order. A *binary relation of order* on a set is *asymmetric* when it is transitive and satisfies the logical property of asymmetry: if two elements of the set are such that the first is in relation to the second, then the second is not in relation to the first.

Partial relations of pre-order constitute another type of relation. A *partial relation* is *of pre-order* when it satisfies only the property of transitivity. The pre-order may also be total.

A *binary partial relation* of *order* is *total* when, for any two elements of the set, the first is in relation to the second, the second is in relation to the first, or both coincide: a relation of order is total when any two elements of the set are comparable according to this relation. In this case, the relation ceases to be strictly partial, being characterized as a relation in the usual sense found in the literature.

A *system* is *ordered* when at least one of the partial relations which characterize it is a relation of order, that is, when there exists at least one 'order' in the system. Thus, a system may be considered ordered under different perspectives, and may have distinct partial and order relations; if one of those relations of order is total, under this perspective it may be considered totally ordered.

It may now be observed that hierarchized systems are ordered ones, and that hierarchies are order relations. When the system is totally hierarchized (given any two elements, one of them is always superior, inferior, or identical to the other), this hierarchy constitutes a relation of total order.

Among a system's partial relations, there may occur partial relations of equivalence. A *binary partial relation* on a set is *of equivalence* when it is reflexive, symmetric, and transitive: a relation is symmetric when, if two elements of the set are such that the first is in relation to the second, then the second is also in relation to the first. Observe that two elements can be equivalent without being identical, according to a partial relation. However, if they are identical they are also equivalent, according to the relation of identity; the relation of identity is a particular case of the relation of equivalence. The set of equivalent elements, according to a certain partial relation of equivalence, constitutes an equivalence class of the system. It may happen that equivalence classes of the system, according to a given partial relation of equivalence, are hierarchized according to another given partial relation of order.

It should be emphasized, then, that in this chapter, a clear distinction between order and organization is made. Organization is an essential characteristic of each system, while order is a particular characteristic of certain organizations and, therefore, of certain systems. Thus, there may be unordered organizations and ordered organizations, which are not totally ordered. There may then exist organized systems whose structures do not possess, among their partial relations, any ordered partial relation.

Partial relations and complexity

Complexity may be characterized by the concept of partial relation. *Complex systems* necessarily present circular partial relations, although their elements are not necessarily numerous. Other systems with many elements, even those with tree-like partial relations, may be considered complicated but not complex (see Le Moigne, 1990; Morin 1990, 1992).

The notions of partial relation and of special types of partial relations allow us to discuss the notion of complexity. An understanding of the concept of complexity may be provided by the relational systemic view. A partial relation, as in the case of binary relations, can be established only between two elements, while a system can only be described in terms of the interrelations among all of its elements. A partial relation derives from a particular characteristic of the elements of the system, while a system is generated by a particular interrelational distribution of its elements. In addition, the interrelations in a system depend on a common reference for the whole set of its elements, and this common reference determines its identity. We observe that a greater variety of elements may provide a greater set of partial relations, and, therefore, greater complexity. Complexity, thus, depends on the amount and variety of elements and partial relations. Finally, complexly organized systems cannot be dismantled or decomposed without being destroyed.

System and organization

Organization, identified by the set of the system's structural and functional characteristics, refers to partial relations and activities or functions of the system, with the capacity to transform, produce, maintain, and generate its behaviors. This characterization includes within it the dynamics underlying the system.

A system's partial structure, the articulated set of partial relations among its elements, may be constituted as an invariant in time. The functioning of a system is conferred by the set of activities of its elements, which conduct the process of transformation, exercising functions in a dynamic way but conditioned by the partial structure. However, the system's dynamics may also derive from a process of structural change.

Organization may be seen as a characteristic of the system founded on the capacity for transforming the diversity of behavior (partial relations and activities) of the different elements in a global unity, but in light of its dynamic and complex behavior it may also be a source of creation of diversity, of capacity, and of structural and functional specificity.

To exercise the role of creation, a system needs to be constituted and to develop in such a way that the field of forces of attraction or cooperation (inclusion, composition, association) among its elements – establishing partial relations and being responsible for organization – prevails over the field of forces of repulsion or competition, represented by antagonisms (exclusion, decomposition, disassociation) and responsible for disorganization. The terms *field of forces of attraction* and *of repulsion* are intuitively used in the sense of the provision of conditions favorable to the existence and to the emergence of specific partial relations.

Fields of forces of attraction and repulsion are inherent to the system as characteristics typical of active organizations. The organization may contain a certain dose of disorganization, which may contribute either to the reduction or to the stimulation of organization. Disorganization may be understood as a limit situation of organization, and vice-versa.

When forces of competition generate the conditions of development prevalent in disorganization, disintegration is propagated. When forces of collaboration generate the conditions of development prevalent in organization, integration is propagated. In both cases, one can say that, at the limit, a systemic crisis materializes as full disorganization (total mobility) or full organization (total immobility). It needs to be emphasized that the origins of forces of cooperation and competition are not only internal to the system, but may also arise from boundary or external elements.

System and finality

Systems may have goals (finalities, purposes, intentions, expectations, meanings). Ascribed by the subject, they need to be understood and explained by the subject, so that a set of characteristic parameters of elements of importation processed by the system can provide a set of characteristic elements of exportation compatible with those goals.

The *effectiveness* of the system expresses its capacity for achieving its goals; *efficiency* expresses the intensity with which it achieves its goals. A system may or may not be effective (its effectiveness may be, respectively, of degree 1 or 0), and may be more or less efficient (its efficiency may vary between degrees 1 and 0). The system's *effectivity* corresponds to the product of the degrees of effectiveness and efficiency, expressing the effectiveness obtained with efficiency.

Activities developed by elements characterize the system's functions, the exercise of which expresses its functionality; a system is a partial functioning structure, with functionality. One should not confuse the terms function, functioning, and functionality.

We cite here two characteristics of systems that are associated with finality:

a   The *teleological* characteristic, which expresses the system's behavior that is directed to a final state representing finality. Finality is distinct from linear causality, in which the effect temporally follows a cause. Finalist behavior happens according to a process which depends essentially on conditions that arise during the development of this process, while in causalist behavior past conditions are determinants of the process;

b   The *equifinalist* characteristic is represented by behavior with a tendency towards a determined final state, originating in different initial states and developing through different paths. A teleological system may be equifinalist, but every equifinalist system is teleological.

System and environment

Environment is everything which is stipulated to be outside of the system's domain. The *environment's universe* is the complement to the system's universe, relative to a certain totality of elements.

In general, a system is not completely isolated from its environment, because everything (matter, energy, information – notions not discussed in this chapter) that goes into or out of the system comes from, passes through, or goes out to the environment; the boundary is where this importation and exportation occurs. However, one can admit the existence of systems which do not interact with the environment; these systems are characterized as isolated or closed to the environment. One can also admit the existence of systems with elements totally sensitive to environmental contingencies; these systems are characterized as open to the environment.

There are three kinds of elements: elements that are internal to the system, those that are external to the system, and boundary elements.

An *element* is *interior* to the system if, and only if, it belongs to the system's universe and there exists at least one partial relation (a component of the system's partial structure) which is maintained by it only with elements from the system's universe. This relation is not maintained with any element from the environment's universe.

An *element* is *exterior* to the system if, and only if, it belongs to the environment's universe and there exists at least one partial relation (a component of the environment's partial structure) which is maintained by it only with elements from the environment's universe.

An *element* is a *boundary* one if, and only if, every partial relation in which it participates is necessarily maintained with elements from the system's universe and with elements from the environment's universe.

The system's *interior* is the set of its internal elements, the *exterior* is the set of its external elements, and the *boundary* is the set of its boundary elements. Boundary elements, therefore, have the task of establishing the system's interrelations to the environment and vice-versa, being responsible for the system's inputs and outputs. The elements of importation and exportation are exactly the boundary elements; they are neither internal nor external. They may belong to the system's universe or to the environment's universe, that is, the boundary elements are determined elements of the union of these two universes. An element is a boundary element precisely because all of its relations are with internal and external elements.

Every boundary element may be external to the system, and in this case, the *system* is said to be *open*. If every boundary element belongs to the universe of the system the *system* is said to be *closed*. However, in none of those cases, even the extreme ones, is the system characterized as isolated or closed to the environment. If the system is open, then the environment is closed, and vice-versa.

A system is *isolated* when it does not maintain any interrelation to the environment. In such cases, a system does not have boundary elements and the boundary is empty. It is very hard to identify isolated systems in nature; isolated systems are usually artificial.

Internal, external, and boundary elements act, interrelating or interacting, under internal and external restrictive conditions of certainties (predicted, determined events) and uncertainties (unpredicted, contingent events), in such a way to allow (or at least facilitate) or impede (or at least hinder) the development of the system's processes. The relations of internal and external elements, in the face of conditions of internal or external certainty and uncertainty, may conduct the system to processes in part predicted or predictable, and to processes in part unpredicted or unpredictable.

The notions of complement, internal, external, and boundary element introduced here derive from the corresponding logical-topological concepts.

Flow and field

As previously mentioned, for a system to develop a flow of activities characteristic of a process, there must be present, as a characteristic of it, a field of forces of influence or catalysis (positive or negative). This field underlies – in a dominant, auxiliary, or co-auxiliary way – the partial relations between elements, the activities developed, and the system's functional and structural alterations. Beyond the flow of relations and activities, there may also occur a flow of structure and functionality, that is, a flow of organization.

Therefore, in a complex system, the existence of many flows is supposed; they are affected by different fields, articulated among themselves, and are responsible for activities; and they also may alter the system's organizational stability.

Change and equilibrium

The description of the *states* of a system allows for the establishment of a perspective from its exterior, and the description of the organization allows for a perspective from its interior.

A system may be in a state of equilibrium and present the characteristic of stability, or it may be in a state of disequilibrium and have the characteristic of instability. In some cases, it may be characterized as having a potential for instability when the system is in a stable state, or a potential for stability when it is in an unstable state.

In a *state of equilibrium*, the system does not transform, and it maintains its organizational characteristics. In a *state of disequilibrium*, it presents alterations in its organizational characteristics.

Organizational changes are a part or consequence of processes that seek survival, reproduction, evolution, or creation in and by the system. *Emergences* are organizational changes, which occur in or derive from the system (with the exception of survival, which is a precondition of the system's existence).

Changes of state can be identified by changes in behaviors of the system's input and output elements (represented by state variables); each new state can be considered a novelty.

Changes in the system may be derived from predetermined activities and be performed by internal or boundary elements; in this case, predictable ones. But they may also incur from non-predetermined activities, and be performed, spontaneously and autonomously, by internal, external, or boundary elements; in this case, unpredictable.

Furthermore, organizational changes may occur continuously or discontinuously, and in an incremental or radical way. Radical changes may occur from a total rupture with the previous organization; in a limit case, a moment considered critical or of systemic crisis, the radical change event may itself constitute a systemic catastrophe.

We can mention two characteristics associated with the maintenance or change of states:

a   The characteristic of *regulation*, manifested by the maintenance of the state of equilibrium and of the system's existence, in the face of internal and external contingencies;

b   The characteristic of *adaptation*, expressed by a change of state in a new state of equilibrium, which guarantees the maintenance of the system's existence in the face of internal and external contingencies.

Mechanisms of regulation and adaptation, which are not exclusive of one another, arise from the dynamic relations within the system and between the system and the environment; through them the system maintains its existence in equilibrium with the environment.

Regulation is a circular process of command and control by means of a causal chain or an informational circuit which performs retroactive monitoring. It allows for the correction of disequilibrium or deviations which could alter the state that is to be maintained (or the goal to be achieved, in the case of systems with a finality). Mechanisms of regulation or adaptation are part of the processes which lead the system to achieve the condition of equilibrium or disequilibrium, and maintenance or change of state. They can act both on functionality and on the structure of the system; thus, there may be, separately or conjointly, functional regulation, structural regulation, functional adaptation, and structural adaptation.

Regulation and adaptation happen through activities of internal or boundary elements. In the case of the autonomous activity of internal elements, with the eventual participation of boundary elements, the mechanisms of control (maintainers of the system's existence) are self-regulation or self-adaptation.

Organization and Self-Organization

The system's *organization* may be considered under formal and informal aspects, which interrelate dynamically in the process of organizational transformation and complement each other, intertwining in the system's constitution (see Bresciani Filho, 1996). *Formal organization* is constituted by a partial structure, predetermined or preconceived by internal and boundary elements to attend to an intended functionality, possibly in the direction of a prefixed finality. However, even without the existence of a prefixed finality, there may be determination when elements possess a low degree of autonomy.

A system's *informal organization* is also constituted by a partial structure, with a corresponding functioning, which is not predetermined. On the contrary, informal organization derives spontaneously from activities of internal elements, and eventually boundary elements, with high levels of autonomy.

Organizational changes, at least some of them, may also be predetermined, preconceived, or planned, through activities of elements from outside, from inside, or from the boundary. They may be spontaneous and a consequence of the autonomous activities of internal elements (and eventually boundary ones), as well as a consequence of the interaction of those autonomous activities with the predetermined ones.

Both informal organization and spontaneous organizational changes may present unexpected, unpredicted, unpredictable, and uncertain properties and behaviors, based on the existence of relations of synergy and deriving from the high levels of freedom in the activities of the elements, from the high sensibilities of the elements to contingencies, from environmental circumstances, or even from casual events.

These notions of formal and informal organization, treated in works which apply a systemic approach to the study of organization and the possibility of self-organizing processes, derive from the study of social systems. However, in this chapter, those concepts are being applied to any subject system without loss of generality.

Self-Organization is characterized as a phenomenon of transformation or of creation of an organization, which results fundamentally from the interaction of predetermined activities (if there are any) with the autonomous and spontaneous activity of internal elements (and, eventually, boundary and external ones) through recurring processes. The spontaneous activity is the result of the existence of a minimal degree of autonomy of the acting elements; recurring processes need to be present for the autonomous elements to integrate into an organization with self-reference (see Debrun, Chapter 1, this volume).

In some cases, one may admit that the organization that emerges in the system, incurring from the phenomenon of Self-Organization, does not have characteristics which allow its qualification as an organizational creation. Such emergent organization may be qualified only as a reproduction or duplication of an organization that already exists or that has existed before. It should be emphasized that processes of reproduction without novelty have received the name of *autopoiesis.*

Predetermined changes may be concurrent or concordant with spontaneous changes and may complement and facilitate them. However, predetermined changes may also be divergent, discordant, and antagonistic to spontaneous changes, and may cause difficulties in their development, contributing to a state of contraposition, contradiction, and conflict in the system.

The influence of autonomous activities of the elements of the environment on the system may be characterized as noise (perturbation or fluctuation) introduced into the system; noise may contribute, in some way, to the occurrence of the phenomenon of Self-Organization (see Atlan, 1979; von Foerster, 1984; Morin, 1977–1991).

Creation, evolution, and Self-Organization

Creation may be the result of transformations conducted by spontaneous and autonomous activities, or from transformations conducted by constitutive and predetermined activities of elements of the system (and eventually, boundary elements); it also may result from the interaction of these two types of transformations. It may be a new product or the result of a process of organizational transformation characterized by the formation of new structures or new functioning. In both cases, creation may be thought of as the emergence of a system.

Creation results from the influence of different factors, particularly those related to degrees of autonomy and to the constitutive nature of the elements of the system (and eventually boundary ones), such as elasticity, plasticity, and, in some cases, the capacity of imagination and conception. It is also important to consider the influence of factors related to the existence of a systemic organization favorable to transformations, or of an environment which is a motivator (promoter, catalyzer, and disturber) of the creation process.

The interrelations between creation and organization may be contained in a process represented by a recurring circle, in which the organization propitiates the performing of creation, and this creation, when performed, propitiates the changing of the organization itself. The recurring cycle may also occur in the interrelation of the system to the environment. The environment propitiates creation in or of the system, and this changes the characteristics of the environment.

The dynamic process of the interrelation of the system with the environment may guarantee the survival (maintenance and renewal), the reproduction, and the evolution of the system. Creation is a process which is not simply identified with this global process, although it may be a part of it. An important condition must be present, which is the necessity of the system to have a reference in relation to itself in order to not mischaracterize itself or lose its own identity. This self-reference may be contained in a memory of the system, which registers past existence represented by a sequence of former states.

The process of *evolution* is characterized as the sequence of states of equilibrium and disequilibrium, manifested in the succession of distinct organizations which arise through the course of the transformation of a system, due to the action of internal, external, and boundary elements. If every organization that arises is considered a novelty, then one can affirm that evolution is a sequence of organizational innovations that may be rightly referred to as *creative evolution*. However, evolution is not only a manifestation of the progress of the system, with the concentration of elements resulting in the construction of a new organization; it may be a process of dispersion of elements or of degradation in the direction of disorganization. Evolution may also be the generator of a diversity of organizations.

Self-Organization and creation may be interrelated through a process represented by a recurrent circle in which Self-Organization propitiates the performing of creation, and creation propitiates the change of the organization in the form of Self-Organization.

A bit of formalism: revisiting the definition of system

For the interested reader, we review the concept of system, making use of a certain logical formalism.

Definition 4. A *system* is a partial structure with functionality that can be denoted by:

$$S = \left\langle D_i, \ R_{ji} \right\rangle^{Fk} i, j, k \in I, J, K,$$

D_i being the universe of the partial structure, each R_{ji} a partial relation on D_i, F the functionality, with I, J, K being the respective variation indexes.

Observation 1. A system, without the possibility of structural or functional alterations, has its universe and functionality constant and can be denoted by:

$$S = \left\langle D, R_i \right\rangle^F i \in I.$$

Final considerations: basic precepts of the systemic approach

As a consequence of the notions, concepts, and definitions presented earlier, a number of conditions and principles must be recognized for a problem to be treated from the systemic perspective:

1 The existence of the system, with an underlying partial structure, constituted by a set of elements and by partial relations between those elements, and with a functionality.
2 The presence of an observer subject, of a complex nature, and with the possibility of belonging or not belonging to the system.
3 The characterization of internal, external, and boundary elements.
4 The identification of partial relations in distinct degrees of complexity.

5   The existence of the properties of synergy, globality, and novelty.
6   The possibility of the system to receive energy, matter, and information from the outside (the environment), to transform it internally, and to transmit it to the outside through its boundary.
7   The identification of a property of the system, called organization, which is responsible for the behavior of the system and is characterized by its partial structure and functionality.
8   The existence of teleological properties or of the equifinality of the system.
9   The appearance of restrictive and disturbing conditions of determined or undetermined characteristics, due to interaction with the environment through the boundary of the system.
10 The necessity of the existence of a field of influence (or of forces) that triggers a flow of activities.
11 The possibility of the maintenance of structural and functional equilibrium, i.e., the maintenance of the system's state in the interactions with the environment, through the mechanism of regulation.
12 The possibility of the presence of the phenomenon of Self-Organization deriving from the interaction of the system's predetermined activities with the autonomous and spontaneous activities of its elements, in interaction with the environment in a recurring process.
13 The possibility of change of state and the emergence of a new state, which characterizes creation or evolution, through the mechanism of structural and functional adaptation.
14 The possibility of transformations through creative processes that could derive from the phenomenon of Self-Organization.

References

Atlan, H. (1979). *Entre le cristal et la fumée. Essai sur l'organization du vivant.* Paris: Seuil.
Bresciani Filho, E. (1996). Organização informal, auto-organização e inovação. In: Debrun, M., Gonzalez, M. E. Q., and Pessoa Jr., O. (Orgs.), *Auto-organização: estudos interdisciplinares* (pp. 365–380). Campinas: CLE/UNICAMP. (Coleção CLE, v. 18).
Bresciani Filho, E., and D'Ottaviano, I. M. L. (2000). Conceitos básicos de sistêmica. In: D'Ottaviano, I. M. L., and Gonzalez, M. E. Q. (Orgs.), *Auto-organização: estudos interdisciplinares* (pp. 283–306). Campinas: CLE/UNICAMP. (Coleção CLE, v. 30).
Bresciani Filho, E., D'Ottaviano, I. M. L., and Milanez, L. F. (2008). Conceitos básicos de sistema dinâmico e térmico. In: Bresciani Filho, E., D'Ottaviano, I. M. L., Gonzalez, M. E. Q., and Souza, G. M. (Orgs.), *Auto-Organização: estudos interdisciplinares* (pp. 19–32). Campinas: CLE/UNICAMP. (Coleção CLE, v. 52).
D'Ottaviano, I. M. L. (2010). On the theory of quasi-truth. In: Agazzi, E., and Di Bernardi, J. (Eds.), *Series Special Issues of Epistemology: Relations between Natural Sciences and Human Sciences-Rélations entre les Sciences Naturelles et les Sciences Humaines* (pp. 325–340). Genova: Tilguer-Genova Publishing Company.
D'Ottaviano, I. M. L., and Bresciani Filho, E. (2004a). Sistêmica, auto-organização e criação. *Multiciência* (vol. 3, pp. 1–23), COCEN/UNICAMP.

———— (2004b) Sistema dinâmico caótico e auto-organização. In: Souza, G. M., D'Ottaviano, I. M. L., and Gonzalez, M. E. Q. (Org.), *Auto-organização: estudos interdisciplinares* (pp. 239–258). Campinas: CLE/UNICAMP. (Coleção CLE, v. 38).

D'Ottaviano, I. M. L., and Hifume, C. (2007). Peircean pragmatic truth and da Costa's quasi-truth. *Studies in Computational Intelligence (SCI)*, 64, pp. 383–398.

Le Moigne, J. L. (1990). *La modélisation des systèmes complexes*. Paris: Dunod.

Mikenberg, I., da Costa, N. C. A., Chuaqui, R. (1986). Pragmatic truth and approximation to truth. *The Journal of Symbolic Logic*, 51(1), pp. 201–221.

Morin, E. (1977–1991). *La Méthode* 1 – *La nature de la nature* (1977); 2 – *La vie de la vie* (1980); 3 – *La connaissance de la connaissance* (1986); 4 – *Les idées* (1991). Paris: Seuil.

———— (1990). *Introduction à la pensée complexe*. Paris: Seuil.

———— (1992). From the concept of system to the paradigm of complexity. *Journal of Social and Evolutionary Systems*, 15(4), pp. 371–385.

Peirce, C. S. (1958). *Collected Papers of Charles Sanders Peirce*, 8 vols. Edited by Hartshorne, C., Weiss, P., and Burks, A. Cambridge, MA: Harvard University Press.

Shoenfield, J. R. (1967). *Mathematical Logic*. New York: Addison Wesley.

Tarski, A. (1944). The semantic conception of truth. *Philosophy and Phenomenological Research*, 4, pp. 13–47.

von Foerster, H. V. (1984). *Observing Systems*. Seaside, CA: Intersystems Pub.

4

THE ROLE OF INFORMATION IN SELF-ORGANIZATION

Alfredo Pereira Jr. and Maria Eunice Quilici Gonzalez

Introduction

According to Debrun (Chapters 1 and 2, this volume), a process of Self-Organization involves three phases; the first two correspond to primary Self-Organization and the third to secondary Self-Organization. The first two phases were extensively analyzed by Debrun (Chapters 1 and 2, this volume), Haken (2000), and Gonzalez et al. (2004), among others; the third, which is the focus of this chapter, has been little discussed in the literature.

In the process of primary Self-Organization, various elements with independent histories interact with each other, sometimes by chance, eventually forming a new system. In such circumstances, abrupt transformations can generate instability that, in turn, may result in the constitution of different systems. In these situations, relations of dependence are created between the distinct elements that, mainly by their own interactions, compose the structure and organization of new systems. As emphasized by Debrun (Chapter 2, this volume), in the first phase of primary Self-Organization ruptures occur in the history of the system. These are, generally, due to fortuitous encounters, or, in the case of the life dynamics of a human being, due to individual or collective decisions that do not necessarily follow from previously established rules. Debrun stresses that chance, understood as the intersection of causal chains with independent histories, plays a key role in this first stage of the process; a new identity begins to be outlined, constituting an inside and an outside face of a fragile system.

The second phase of the process of primary Self-Organization is, according to Debrun, marked by an "endogenization" through which distinctions between the internal and external aspects of a system are accentuated. In this way, chance, which plays a fundamental role in the initial state of a self-organizing process, is diminished in this phase; the possible histories that come into being with the members of the new system are forgotten, they are "neutralized, redefined, or subordinated to the global movement of the process" (Debrun, Chapter 2, this volume). At this stage, the new system acquires relative stability of its own. Finally, in the third stage of the Self-Organization process, the crystallization of a form or identity might occur, indicating that an organization has been achieved or is to be achieved.

In the present chapter, we propose a way of understanding the fundamental third phase of Self-Organization that characterizes an essential aspect of the dynamics of secondary Self-Organization. We consider that different origins and causes of secondary Self-Organization are possible (including the possibility of engineering conducted partly by an external agent) as long as they do not detract from the spontaneous conditions of Self-Organization processes. The concepts presented here to explain aspects of self-organizing processes are limited to those that involve information. We do not intend to delve deeper into the discussion of these aspects in themselves (the concept of information was previously discussed in Pereira Jr. and Gonzalez, 1995), but mainly to apply them to the study of Self-Organization, distinguishing different types of processes and modalities of information that play a central role in the understanding of secondary Self-Organization.

In the processes of secondary Self-Organization, as noted previously, the interaction among components of a system can give rise to the emergence of mutual dependence relationships and functions that incorporate traces or memories of their constitutive form, which we call *patterns*. Pattern generation can occur both in the causal dimension (here understood as efficient cause), in which the dependencies among the parts of the system manifest themselves mechanistically or deterministically, as well as in the information dimension. This contrast allows for greater flexibility between the components of complex systems.

In the following section, we propose to distinguish between the causal and informational dimensions present in complex systems, arguing that the informational dimension is of central importance for understanding the dynamics of secondary Self-Organization processes. According to this perspective, organizational changes result from processes in which different patterns, internal and external to a system, intersect and generate new patterns that delimit the system's identity. We then propose a distinction between four modes of information, which we call *structural, environmental, contextual,* and *anticipatory.* On the basis of these four modes of information, we illustrate the nonlinear dynamics of information flows present in different systemic modalities that characterize processes of secondary Self-Organization. Modeling informational dynamics of this kind requires inter- and/or trans-disciplinary tools that are more advanced than those allowed by our current knowledge. Therefore, we will limit ourselves to the presentation of a theoretical framework, distinguishing different types of information and the general categories in which secondary self-organized processes can be framed, leaving for future investigations a more detailed approach to their possible modeling.

Causal and informational dimensions

In a previous work, Pereira Jr. and Gonzalez (1995) proposed the hypothesis that informational relations are not opposed to causal relations, but may overlap in certain situations, constituting a kind of "second order causality" in which established organizational patterns become correlated.

Accordingly, informational relationships can be found among organizational patterns correlated in different systems.

In the present chapter, we draw attention to an important distinction to be made between the notions of causality and information in self-organizing systems. Causal relations (restricted here to efficient cause) may have a deterministic character, in the sense that the laws and principles of nature require (in a necessary way) that the occurrence of a state of affairs in one part of nature produces another state of affairs in another part of nature (or in the same part at a subsequent time). Such a necessary imposition does not always apply to causality, given the existence of "statistical causality" (Salmon, 1984) in which the occurrence of an event only increases the likelihood of a new event; a classic example of this is the relationship between smoking and lung cancer. As is well known, there are many cases of people who smoke intensively throughout their lives and do not contract the disease, although it is known from other and more numerous cases that smoking is a serious risk factor in the triggering of the cancerous process.

In statistical causality, there is the intervention of a component of chance that erodes the deterministic character of the strict obedience to laws and principles of nature by introducing an aspect of contingency. This may derive from the ultimate nature of physical laws and principles themselves and/or the initial and contour conditions according to which the process in question is established. Considering the example just presented, it may be argued that "genetic factors" (considered as initial conditions) and "environmental factors" (such as air quality, type of food, and physical activity, considered as contour conditions) could be decisive in triggering (or not triggering) the carcinogenic process in the smoker.

In informational processes, it is not only the introduction of a contingency factor that matters but also the possibility of communication among the components of a system and between the system and its external environment. Such communication is achieved by means of signals that can trigger several possible results in the receivers. In the biological context (as we shall discuss later), these signals can have a meaning that is developed in a process of co-evolution among living systems and their environments. An example, often cited by Debrun (personal communication), is that of traffic signals: a red light indicates that the driver must stop the vehicle, but at the same time he is not deprived of the freedom to cross the road while the red light is on (as justified, for example, in situations where it is necessary to take someone to the hospital or where there is a risk of a thief approaching a stationary car). We understand, therefore, that the informational dimension differs from the causal dimension (both deterministic and probabilistic) by introducing a choice factor; namely, the establishment of communication between two systems by means of signals emitted and received, and by which alternatives of action are opened up for the receiving system. In addition, the presence of noise, understood as various kinds of interference in the transmission/receiving process, can contribute to the disorganization of system functions, eliminating or generating new patterns of actions. Examples of the distinction between the causal and the informational dimensions are brilliantly presented by Gibson (1979), through the study of the concept of *affordance*.

Roughly speaking, affordances are relations of mutual dependence established between organisms and their environment that give place to dispositions indicative of possibilities of actions (Turvey, 1992). Thus, in the terrestrial environment, Gibson investigates the conditions of luminosity that indicate possibilities of action for different types of human walking: under suitable light conditions, flat or slightly sloping terrains afford a stable kind of footstep, while very irregular or sandy surfaces require a different type of movement that has to be performed carefully in order to maintain balance. Under suitable conditions of luminosity, temperature, air humidity, etc., organisms move around and avoid accidental shocks, thanks in large part to communication among the various elements of the environment and including the organism itself. The dynamics of self-organized locomotion might be interrupted by the presence of noise when, for example, a dense cloud of smoke enters the environment and alters the respiratory, visual, and olfactory functioning of organisms, preventing detection of the usual affordances. In such circumstances, organisms (if they survive) will have to find or create new informational patterns that can fulfill the role of affordances in their various planes of corporeality. Given the relational, evolutionary, and historical nature of affordances, constructed through the action of organisms in their environment, understanding them presupposes an ecological-informational perspective that is distinct from a strictly physical one. In this perspective, the physical universe becomes an ecological (biologically meaningful) universe of action, also known as an ecosystem or a *niche*.

Thanks to the self-organized action of organisms, which carry their co-evolving history within themselves, affordances are formed, creating an ecological dimension to their niches. This ecological dimension of the actions of living beings involves not only physical stimuli and information devoid of meaning, but also meaningful information systems loaded with co-evolutionary histories. Within these pregnant systems of histories, physical stimuli become meaningful due to their functionality in the establishment of dwellings for numerous forms of life; it is in this informational ecological system that affordances are central.

The ecological dimension of affordances illustrates our hypothesis about the nature of informational relations, characterized now as a kind of "second order causality": they stand at the dynamic crossroads of emerging environmental patterns of (first order) causal relations and patterns of actions that are evolutionarily constituted by the immense variety of organisms. In this circumstance, the traditional separation between deterministic or probabilistic causal processes, on the one hand, and meaningful informational processes on the other, becomes vague.

We assume that in their niches, complex systems embody, to a greater or lesser degree, the dimension of deterministic causality (contingent on probabilistic causality), as well as the dimension of the possibility of actions, which is present in their communication processes.

In short, we consider that causality (in its two modalities) and information constitute complementary dimensions of the processes involved in actions. Consequently, in constructing explanatory models of the temporal evolution of complex systems possessing a niche, the study of information complements the study of causality, especially with respect to the emergent properties of self-organizing systems. The informational dimension acquires central importance for living systems, since it is from the dynamics of flows of information that meaningful informational patterns (affordances) are established, enabling the self-organizing system to guide its own evolution (thus, justifying the use of the prefix "self-"). In contrast, systems considered "hetero-organized" are those that have their evolution guided mainly by causal relations and/or external agents. This reasoning leads us to conjecture that if systems possessed only the primary dimension of (efficient) causality, and were devoid of the informational dimension, then they would be guided exclusively by the forces of necessity and chance, lacking the resources to establish, in an immanent way, their own goals and the strategies required to achieve them.

Organizational complexity incorporated in living systems

To exemplify some aspects of the process of the secondary Self-Organization of complex systems, we will initially refer to living systems (for previous discussions on this topic, see Pereira Jr. et al., 1996, 2004). These systems are usually characterized as having several organizational dimensions, each containing components that interact among themselves and forming functional units that allow the emergence of properties not observed in their components in isolation. In the case of living beings, we have the following superimposed organizational levels: molecules that interact with each other to form cells; cells that interact with each other to form tissues; tissues that interact with each other to form organs; organs that interact with each other to form physiological systems; physiological systems that interact with each other to form organisms; organisms that interact with each other to form populations; and populations that interact with each other to form ecosystems.

Studies of the organizational layers of complex self-organizing systems require distinctive methodological characteristics for the identification of different types of processes and information modalities. We do not ignore the possibility that various forms of organization might incorporate an ontological aspect as well, but we do not think it would be convenient to present extensive arguments on this subject here (for a detailed study, see Gonzalez et al., 2004).

Considering that studies of living systems require different resources for the understanding of their innumerable organizational dimensions, we propose to discuss the following (sub)processes in order to investigate their organizational layers:

a   Internal "horizontal" processes, occurring among elements of the same organizational domain;
b   Internal "vertical" processes, involving interactions among organizational levels;
c   Interaction of a system with its environment, through which parts of both produce effects that spread through the system, which then becomes understood as a whole.

In weakly hierarchical systems (such as those that self-organize), two types of vertical processes can be distinguished:

a   Upward causation, in which elements of "lower" hierarchical layers influence or enable the emergence of informational patterns, also known as *order parameters*, in the "higher" hierarchical layers (Haken, 2000);
b   Downward causation, in which elements of hierarchical "higher" layers, constituting order parameters, influence or restrict (enslave) the behavior of the elements of the "lower" layers (Haken, 2000).

In horizontal processes, we consider that the elements present in each layer contain their own causal determinations. However, when they meet and begin to interact systematically, new functional patterns can emerge which, through a process of upward causation, influence elements at the higher organizational layer. These elements, in turn, provide feedback to the components of the lower layers, via downward causation, generating structural modifications in their constitutive systems. Thus, for example, genes interact with each other, determining the production of a pool of proteins that engender cellular functions. In turn, these functions provide feedback in the regulation of gene expression, altering the profile of the proteins produced and shaping their functions, and so on (Pereira Jr. et al., 1996).

The aforementioned hypotheses concerning the dynamics of Self-Organization, proposed by Debrun (Chapters 1 and 2, this volume) contribute to understanding the nature of secondary Self-Organization processes and their links to systemic evolution. In particular, the notion of *adjustment,* characterized as a type of learning process involved in the establishment of mutual relationships, has important theoretical relevance. Information about the learning resources involved in adjustments might be necessary in order to decrease encounters that involve clashes of forces (often antagonistic) between elements, threatening the stability of the system. Debrun (Chapter 1, this volume) provides the example of an old couple (typical of many couples), who, from the accumulated results of many disputes, have learned how to deal with information about each other. This example illustrates the complexity of sub-processes involved in the embedding of the informational patterns inherent in the constitution, evolution, and learned stability of secondary self-organized systems.

In short, complex self-organizing systems may involve sub-processes containing both the causal-deterministic dimension and the informational dimension. We attribute special value to the learning/informational dimension of the secondary process of Self-Organization, as it enables interactions between various components and functional units of a system without prejudicing the system's "degrees of freedom". On the basis of this dynamics, global interference patterns can be formed that guide systemic evolution in an immanent way, giving rise to new patterns of organization. In the next section, we propose a distinction among four modes of information that seem to play central roles in the processes of biological Self-Organization in general and, in particular, in processes of Self-Organization in human systems.

Modalities of information flow

As a working hypothesis, we postulate four types of information operating in processes of secondary Self-Organization (they are not exclusive; in certain cases, there may be intersections between the specified domains):

a   Structural information: An inheritance of primary Self-Organization that has already been acquired by systems; it is the set of standards necessary for the reproduction of the organization of systems, such as DNA for living beings, or programs that enable computational simulations of a neural network;

b   Environmental information: The set of constitutive patterns of niches, or of physical, biological, and cultural environments; this type of information operates at the level of "horizontal" interactions between organisms and their environment, and it can trigger "vertical" processes within the system which are characterized as adaptive (when the system conforms to an external situation) and/or proactive (when the system seeks to adjust external conditions to its own determinations);

c   Contextual information: The set of patterns of information available to subsystems of complex systems, which concern the properties and/or states of other subsystems. The concept of contextual information can be illustrated by several situations involving individual and/or collective organisms. Thus, for example, in the various layers of organization of living systems, there are signal exchanges (chemical and electrical) that alter or modulate the activity of a cell, in accord with the acquisition of information about the activity of other cells. Circular contextual information among the components of a given system may result from its physical and biological history, and may often be difficult for an external observer to apprehend (consider, for example, the extremely complex informational interactions that occur between calcium ions and nerve or muscle cells). This type of information can trigger collective phenomena whose consequences can affect other organizational levels;

d   Anticipatory information: This type of information refers to elements that each component of the system has in relation to its goals in specific circumstances. In agreement with Debrun (Chapters 1 and 2, this volume), we consider that goals are not given a priori in self-organizing processes, but rather are established by the system itself during the course of its consolidation. Anticipatory information that helps in the organization of biological systems can be of fundamental importance for supporting adaptive processes such as survival and reproduction. When this type of information is available to various organizational layers, dispositional states can be generated that play a central role in systemic evolution. Such dispositional states are characterized by potential relationships that, under appropriate conditions, enable the implementation or materialization of events (Ryle, 1949). Although little is known about the ontological status of dispositional states, examples such as the solubility of salts in water, the fragility of glass that breaks due to the impact of a given force, the growth of certain microorganisms in specific light environments, and the tendency to consume products widely promulgated by the media, among others, illustrate such states in the areas of physics, chemistry, biology, and sociology.

An understanding of the dynamics of information types applied in human action and personal identity is of great interest to current research on Self-Organization. We believe that the emergence of new characteristics in the self-organized biological, psychological, and social systems that, among other things, characterize the identity of a person, might have its main origin in the Self-Organization process itself, through spontaneous interactions among the constitutive elements. As suggested by Debrun, "The self-organizer subject remains 'within itself' during the restructuring operations. It performs a task of itself on itself"; and he considers that:

> The initial conditions (the very existence of the organism, the biological, social, and cultural context in which it performs) as well as the interchange with the environment (energetic, material, informational, symbolic, or other) play an important role, but only a supporting one, either through challenges (noise or threatening competition, for instance) or through actual or potential resources offered, or yet through goals suggested as a response to challenges.
>
> *(Debrun, Chapter 2, this volume)*

The study of processes of Self-Organization that occur in a person's life system (given its immense complexity) is only the beginning. Pereira Jr. et al. (2002) identified domains of human activity that involve interrelated subsystems, each with their own functions (family, body, work, leisure, sociability, and others).

One hypothesis to be investigated is that the interactions among remarkable events of an individual's mental life could be associated with a search for satisfaction and be related to reference patterns incorporated in his/her life history. This search would enhance the interactive dynamics among subsystems of an individual's physical/mental life. From this interaction, learning processes involved in secondary Self-Organization could allow new global systemic patterns to emerge, corresponding to mental health or to mental crises, and leading to mental disorder or to harmonious patterns of existence.

Types of self-organizing dynamics

In complex systems, the four types of information proposed previously interact in self-organizing processes, often generating nonlinear effects disproportionate to the relative importance of their isolated elements. From an analysis of the forms of interaction between subsystems outlined in the previous sections, three types of informational interactions can be considered. These are similar to the types of games identified by von Neumann and Morgenstern (1944), and they characterize the dynamics of secondary Self-Organization:

a   Stationary: A dynamic in which the actions of several subsystems pursuing different goals end up counteracting one another in the totality of the self-organized system. This generates a stable organization pattern, but one without manifestations of creativity.

b   Conflict: A dynamic in which conflicts between subsystems are damaging to the system as a whole, which as a result undergoes transformations that threaten its stability. Here, among other important possibilities, a failure might occur in the communication of contextual information, since the various subsystems have not agreed to avoid such a conflict. This view pervades certain interpretations of the Darwinian theory of biological evolution, where an emphasis is placed on the competition for scarce resources ("struggle for life") as the engine of the transformations that leads to the extinction of some species and the emergence of others (see Pereira Jr. et al., 2004).

c   Cooperative: A dynamic in which cooperation between subsystems becomes constructive for the system as a whole; this allows for interpretations of evolutionary theory in which various forms of life are actively engaged in adaptation to the environment, in order to fulfill their needs by associating themselves in different ways with other species that are also seeking survival and satisfaction of needs (Pereira Jr. et al., 2004). In the cooperative dynamic, contextual information (allowing adjustments between subsystems), environmental information (enabling active adaptations of systems to their environment), and anticipatory information (helping their goals to be achieved) can adjust themselves, helping with the maintenance of the system's stability.

From this outline analysis, we can re-elaborate the concept of *dynamic self-regulation*, which seems to improve on the classic concept of homeostasis as an expression of the processes of Self-Organization in biological systems. Dynamic self-regulation is characterized by the nonlinearity emergent from various information patterns, and by the flexibility that comes from the complexity generated by the intersection of the fields of possibilities. Thus, for example, from the moment a component of a system rectifies its anticipatory information, it becomes available for adjustments to its functions, something which would not be permissible if it persisted in the previous pattern.

Final comments

Our aim in this chapter is to investigate the hypothesis that processes of secondary Self-Organization involve changeable relations of mutual dependence that we call *informational relations*. We argue that such relationships allow for a variety of adjustments between the components of systems originally formed in the processes of primary Self-Organization. The informational dimension of the processes of secondary Self-Organization is differentiated from the linear causal dimension by means of the indication of organizational layers, which may be constituted by any of the four modalities of information: structural, environmental, contextual information, and anticipatory.

We also argue that the dynamics of secondary Self-Organization can be expressed in terms of cooperative, stationary, and conflicting informational processes from which new organizational patterns can emerge. The ideas presented in this chapter are only a brief outline of a largely unexplored field of research concerning adjustment processes characteristic of embedded embodied action systems, processes which involve learning in the development of secondary Self-Organization. We believe that future interdisciplinary research on this topic could help us to understand aspects of the complex dynamics of life.

Acknowledgments

The authors thank FAPESP Process N. 16/50256-0 and CNPq for supporting this research, and colleagues from UNICAMP and UNESP for discussion and constructive criticisms.

References

Gibson, J. J. (1979). *The Ecological Approach to Visual Perception*. Boston, MA: Houghton-Mifflin Company.
Gonzalez, M. E. Q., Nascimento, T. C. A., and Haselager, W. G. F. (2004). Informação e conhecimento: notas para uma taxonomia da informação. In: Ferreira, A., Gonzalez, M. E. Q., and Coelho, J. G. (Orgs.), *Encontro com as Ciências Cognitivas* (vol. 4, pp. 195–220). São Paulo: Cultura Acadêmica.
Haken, H. (2000). *Information and Self-Organization*. Berlin: Springer Verlag.
Pereira Jr., A. et al. (2004). Evolução biológica e auto-organização: propostas teóricas e discussão de dois casos empíricos. In: Souza, G. M., D'Ottaviano, I. M. L., and

Gonzalez, M. E. Q. (Orgs.), *Auto-Organização: Estudos Interdisciplinares* (pp. 21–72). Campinas: UNICAMP, Centro de Lógica, Epistemologia e História da Ciência. (Coleção CLE, v. 38).

Pereira Jr., A., and Gonzalez, M. E. Q. (1995). Informação, organização e linguagem. In: Évora, F. R. R. (Org.), *Espaço e Tempo* (pp. 255–290). Campinas: UNICAMP, Centro de Lógica e Epistemologia. (Coleção CLE, v. 15).

Pereira Jr., A., Guimarães, R., Chaves Jr., J. C. (1996). Auto-organização na biologia: nível ontogenético. In: Debrun, M., Gonzalez, M. E. Q., and Pessoa Jr., O. (Orgs.), *Auto Organização: Estudos Interdisciplinares* (pp. 239–269). Campinas: UNICAMP, Centro de Lógica, Epistemologia e História da Ciência. (Coleção CLE, v. 18).

Pereira Jr., A., Lussi, I. A. O., and Pereira, M. A. O. (2002). Mente. In: Parentoni, R., and Mari, H. (Orgs.), *Universos do Conhecimento* (pp. 201–219). Belo Horizonte: Faculdade de Letras da UFMG.

Ryle, G. (1949). *The Concept of Mind*. London: Hutchinson.

Salmon, W. (1984). *Scientific Explanation and the Causal Structure of the World*. Princeton, NJ: Princeton University Press.

Turvey, M. (1992). Affordances and prospective control: an outline of the ontology. *Ecological Psychology*, 4, 173–187.

von Neumann, J., and Morgenstern, O. (1944). *Theory of Games and Economic Behavior*. Princeton, NJ: Princeton University Press.

5

ON THE SELF-ORGANIZING OF REALITY-TOTALITY AS LIVING KNOWLEDGE

Ricardo Pereira Tassinari

Introduction

How is it possible to introduce a notion of Reality as Totality that is consonant with the continuous development of contemporary science? To answer this question, we introduce in this chapter the notion of *Reality-Totality*. We argue in favor of the view that Reality as a Totality may be conceived as active and living Knowledge: the self-exposing Idea that self-exposes itself to us by a self-organizing process of which our knowledge process itself is part. The view we argue for is of a metaphysical nature, and it elaborates on the methodological character of the study of the Self-Organization. We show that this philosophical view emerges from some general reflections on the constitution of scientific knowledge, providing elements that make possible the structuring and coordination of various scientific contents and methods.

Our exposition begins with a discussion of the epistemological analysis of Gilles-Gaston Granger (1920–2016), and the genetic epistemology and psychology of Jean Piaget (1896–1980) and his coworkers. We then proceed to an analysis that leads us to the notion of Reality-Totality and its properties. At the conclusion of the chapter, it will be seen that the view presented here is a new form of absolute speculative idealism, close to the philosophical view of Georg W. F. Hegel (1770–1831).

Scientific and philosophical knowledge according to Gilles-Gaston Granger

We will begin by assuming with Granger that scientific knowledge of the empirical world is mainly characterized by the construction of models, and that there are limitations on these kinds of constructions when they are related to human facts

(Granger, 1988, p. 12; 1992, p. 14; e 1994, p. 245; 1995, p. 70). The main limitation is related to the singularity and multiplicity of the significations existent in human facts. These singularities and multiplicities form a limit to the construction of models in the following sense: when we construct such models, we necessarily abstract from certain aspects and qualities of the human facts that in other contexts influence human behavior; thus, in these other contexts, the constructed model does not completely explain behavior; models of this kind are, therefore, necessarily incomplete with regard to all possible kinds of human behavior. It is true that a new model can be constructed to explain some aspect or property that the earlier model did not explain; but in the construction of this new model, we will again make abstractions of aspects and qualities that in other contexts will influence behavior, and the new model will end up not being complete. The limit of this process of construction of models of human behavior is the world as it is lived singularly by us *here and now*; thus, for all proposed models there is a meaning for human beings that is not included in the explanations these models provide.

As Granger (1995, pp. 85–86; my translation) points out:

> The unique but radical obstacle [to scientific knowledge] seems to me to be the *individual* reality of events and beings. Scientific knowledge is fully exercised when it can neutralize this individuation without seriously altering its object, as usually happens in the natural sciences. The fundamental obstacle is evidently in the nature of the phenomena of human behavior, which carry a load of *significations* that resist their simple transformation into *objects* [models], that is to say into abstract schemas that are logically and mathematically manipulable. A feeling, a collective reaction or a fact of language seems hardly to be reduced to such abstract schemas.

The solution to the limitation on the knowledge of human facts by models is to consider the model as a partial representation of a limit never attained. As Granger (1995, p. 117) emphasizes, in the case of human facts, science strives to increasingly encompass the individual in networks of concepts, without ever hoping to attain this. Therefore, the question is not to reduce them but to represent them, albeit partially, in systems of concepts.

Here it is important to highlight Granger's distinction between scientific knowledge and philosophical knowledge. According to Granger, *philosophical knowledge* is relative to what he calls *metaconcepts* "that do not apply directly to experiences, but to the representations of experience" (Granger, 1995, p. 46), and depends on a set of interpretative rules of the lived Reality established from the originative decisions of each philosopher. Such metaconcepts and interpretative rules define what Granger calls *factum* (in opposition to facts represented by models and subject to verification).[1] In this regard, Granger states: "*We meet then [in the philosophical metatheory] originative decisions that it [the philosophical metatheory] proposes to orient in the organization of the senses of the living*" (Granger, 1988, p. 259).

In this chapter, we explain some of our *originative decisions* related to scientific knowledge and its development, and introduce a notion of Reality as Totality as living and active Knowledge, and as the Idea that self-exposes to us by a self-organizing process. One of these originative decisions is that the consequences of the principles (*interpretative rules* in Granger's language) of our interpretation should not be in contradiction with established facts of the special sciences, and, in particular, with facts about the process of knowledge established by genetic psychology.[2]

From this, it follows that a complete overview of Reality as Totality cannot be constructed by a single model without the expectation that this conception will be refuted or contested by a model that is more explanatory, or by the possibility of the adoption of other metaconcepts and interpretative rules for the philosophical interpretation of Reality.

Note that, here, we admit the existence of various forms of interpretation of Reality (and Reality-Totality) on account of the various possibilities for the adoption of principles by originative decisions. Thus, we consider our proposal here merely one among various possible interpretations. Ours, however, is an interpretation that allows coordination with all others, since, for us, all interpretations aim to expose Reality-Totality for themselves, even if they don't admit it.

The capacity of representing according to genetic epistemology and psychology

In the construction of the necessary structures for knowledge, Piaget and Inhelder (1966a) identify the appearance of the *semiotic function* that consists of being able to represent something (any *signified*: an object, event, conceptual schema, etc.) by means of a differentiated *signifier*, and serving only for that representation. Piaget, in accordance with Saussure (1966), makes a distinction between two (non-exclusive) groups of signifiers that are distinguished by how they signify: the symbol and the sign. The *symbol* is motivated (in the sense that it in some way resembles its signifier) and individual (in the sense that its resemblance is established by the subject itself in his or her action and is not just received from others). Examples of symbols are imitation, design, and mental imagery, which according to Piaget and his coworkers consist of the internalization of imitations (such as a mental image of one's backyard or of childhood schoolyards that today seem to us smaller than before[3]).

The *sign*, of which words are the most characteristic example, is collective and arbitrary (e.g., the English word *water* differs from *água* in Portuguese or *Wasser* in German, etc.), in contrast to the symbol's characteristics of individuality and motivation. As Piaget states:

> The *symbol* and the *sign* are the signifiers of abstract meanings, such as those which involve representation. A *symbol* is an image evoked mentally or a

material object intentionally chosen to designate a class of actions or objects. So it is that the mental image of a tree symbolizes in the mind trees
in general, a particular tree which the individual remembers, or a certain
action pertaining to trees, etc.

(Piaget, 1952, p. 191)

The sign, moreover, is a collective symbol, and consequently arbitrary. It also
makes its appearance in the second year, with the beginning of language and
doubtless in synchrony with the formation of the symbol. Symbol and sign are
the two poles, individual and social, of the same elaboration of meanings.

Piaget considers another type of signifier in which the signifier is not differentiated from its signified: the *index* (or *indication*; see Piaget, 1952, pp. 191–196, and
Piaget and Inhelder, 1966a, p. 42). He calls *signals* the indices that are part of an
artificial situation (as, for example, the experiment by Pavlov in which the saliva-
tion of a dog was associated with a sound of a bell; in this case, the sound of the
bell was a signal of food). Among all of Piaget's types of signifiers, we are most
directly interested in the sign, whose uses (combined with symbols, indexes,
signals, and schemes of action) make us capable of elaborating the knowledge
expressed in theories and models.

The principle of designation of Reality-Totality and the Idea

Based on genetic epistemology and psychology, we can very generally say that
when we have sufficient detailed knowledge about the possible actions of the
objects of our Reality, we proceed naturally to the construction of models and
theories. In this context, we agree with Granger (1988, p. 13; 1992, p. 14; 1994,
p. 245) that a model is a system of signs and operations[4] on them, that we use to
represent the objects of our Reality and our actions on it. Hence, operating on
signs attached to possible actions, we can predict new experimental possible facts
(directly related to *virtual facts*; see Granger, 1992; 1995, p. 49). In addition, we
can explain them by showing how the objects of the domain of study are related
between them. Furthermore, we can explain, based on these relations, we can
deduce particular relations that occur in a particular experiment, which leads to
the process of verification, as explained by Granger (1992). Thus, we have the
following schema:

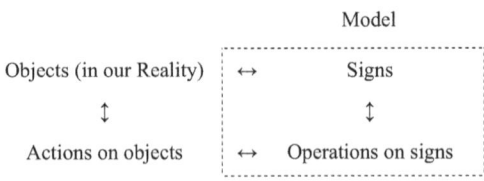

For example, we can consider here a simple model of distances in space: the Pythagorean Theorem. According to the theorem, in a right triangle we have the relation $a^2 = b^2 + c^2$, in which a is the measure of the *hypotenuse* and b and c are the measures of the *catheti*. If the theorem is considered a property of physical space, then it is a relation between the results of the lengths of the *hypotenuse* and the *catheti*. Note that the arithmetical operations expressed in it also indicate possible operations on signs attached to the actions of measurements; we can, therefore, admit that models and theories express Reality.

In response to the initial and central question of this chapter (about the possibility of a notion of Reality as Totality that is consonant with the continuous development of contemporary science), insofar as access to Reality is through signs (of a model or of natural language), we can conceive the Totality as: *all that we can in principle designate by signs.* This notion of Totality can be considered here as methodological, in the sense that it is homogeneous with the construction of models and with knowledge by signs, inasmuch as it is defined on the basis of sign designations and the structures formed by them. The following principle introduces the notion of *Reality-Totality* and summarizes this conception:

> Principle of Designation of Reality-Totality: *What the sign* Reality-Totality *designates is the system of all that we can in principle designate by signs.*

If, following German idealism, we consider the sign *Idea* to designate the system of all our comprehensions of Reality-Totality, then we can introduce in this context the following operational definition of Idea:

> Operational definition of Idea: *Idea is the system of the totality of all things we can designate by signs.*

We, thus, assume here, by the Principle of Designation of Reality-Totality, the following equality:

> *Reality-Totality* = *Idea.*

Notice that what is designated by symbols and indexes can also in principle be designated by signs. Therefore, what is designated by symbols and indexes is also part of the Idea. All that can be known is within the scope of Idea, including what is possible to be known by scientific means. But it is not possible to expose *immediately and completely* for us what Reality-Totality or the Idea is; what the sign *Reality-Totality* designates can only be exposed partially and gradually. Therefore, Reality-Totality or the Idea will be considered simultaneously as the partially exposed result of the gradual process that exposes it, and also as the process itself. In the following section, we will see how this leads to a self-organizing process.

Principle of conceptual characterization of Reality (and Reality-Totality)

If we now assume the Principle of Designation of Reality-Totality, and along with it the idea that contemporary science is a model-maker necessary for our comprehension of Reality (and Reality-Totality[5]), then we can also assume the following principle:

> Principle of Conceptual Characterization of Reality (and Reality-Totality): *What the sign Reality (and Reality-Totality) expresses depends on the construction of models.*

Notice that the signifieds and the significations of some signs of scientific models or theories are not restricted to the operational significations inside them, but overflow them and are evidently anchored in our *here and now daily life*, which, as noted earlier, is a limit never attained by the construction of models. Recalling Granger, we can say that here it is about *representing* the facts in conceptual schemas and models and not *reducing* the facts to models.

Adopting then the Principle of Designation of Reality-Totality and the Principle of Conceptual Characterization of Reality, some consequences may be derived:

a   Experiments (and the possible actions that are necessary for them) are interpreted inside a model or theory, always leading to *theoretically charged* interpretations.

b   The objects in the experiments (on which we perform actions) are, therefore, defined by models or theories, and by their structures and their relations with experimental methods.

c   In this sense, the structures expressed by the models and theories constitute *part of* the structure of Reality-Totality.

Notice that the Principle of Designation of Reality-Totality and the Principle of Conceptual Characterization of Reality-Totality imply that it makes no sense to talk about things that are outside of what we designate by the sign *Reality-Totality*, and that, therefore, there is no sense in the idea of a thing-in-itself that is not in relation to the process of knowing. Also note that, in this case, the general process of knowledge has its contents (and also its form, as we will see in the next section) related to the logical and ontological Idea, which precedes any process of knowledge of a particular subject.

The principle of the Ideality of Reality-Totality

We will then assume the following principle that is made up of the conjunction of three assertions.

Principle of the Ideality of Reality-Totality: *The Principle of Conceptual Characterization of Reality-Totality; It is impossible to construct a unique complete model of Reality-Totality; An uninterrupted construction of models is therefore necessary in order to know what Reality-Totality is.*

This is called *Principle of Ideality of Reality-Totality* because it asserts that Reality-Totality depends on a conceptual characterization that exists for itself as a limit of our knowledge of it.

Insofar as this totality closes on itself, we must admit that the process of knowledge carried out here is also in Reality-Totality, since there are beings (us) that belong to it and carry out this process. Therefore, Reality-Totality exposes itself for us by this very process, and what Reality-Totality is must be identified simultaneously with the result of this process and with this process itself. In this sense, the sign *Reality-Totality* designates:

a   Something in which there is a proper process of knowing itself and the exposition of itself, or better, the self-exposition of itself;

b   Something that is identified by each subject in each moment with the product of its process of knowledge of Reality-Totality;

c   Something that is made more complex in each moment by this process that is exposed of itself;

d   Something under the Principle of Ideality of Reality-Totality; and, therefore, also

e   Something under the Principle of Conceptual Characterization of Reality-Totality;

f   Something for which there is no unique complete model to express;

g   Something for which there is the necessity of an *uninterrupted* construction of models.

These characteristics allow us to see the process of the self-exposition of Reality-Totality as a self-organizing process, as we will see in the next section.

The self-organizing of Reality-Totality

From the qualities listed in the last section, we can characterize this process of knowledge and self-exposition of Reality-Totality as a *self-organizing*, as defined by Michel Debrun (Chapter 1, this volume). Note first, however, that the self-organizing process of self-exposition of Reality-Totality is the general process that involves all self-organizing processes of knowledge about Reality-Totality on the part of the subjects. In this context, it is possible to apply to these process Debrun's general definition of Self-Organization, *"an organization or 'form' that is self-organized when it produces itself"* (Debrun, Chapter 1, this volume). According to these principles, the form of the knowledge process is part of the form of the process of Reality-Totality and it *self*-exposes *itself*. In this case, there is a *self-constituted* system, and so it is, in fact, a secondary form of Self-Organization, according to Debrun's classification:

> When there is an external plurality – which goes from dissociated elements to the constitution of a form – we can say that we are dealing with *primary* Self-Organization... When, on the other hand, it is a matter of the *self-complexification* of a self-constituted organism (or, more generally, of a system), we are dealing with *secondary* Self-Organization.
> *(Debrun, Chapter 2, this volume)*

Lastly, we can apply Debrun's definition of secondary self-organizing process. According to Debrun (Chapter 1, this volume): "*Secondary Self-Organization occurs when, in a learning process (corporal, intellectual, existential, or other), the interaction occurs between the parts ('mental parts' and/or 'corporal parts') of an organism... under the hegemonic, but not dominative, guidance of this organism's 'subject-face'*". Clearly, the process analyzed here is a case of intellectual and corporal learning, since it expresses itself by the increased complexity of the subject's knowledge. The subject is the *subject-face* of the organism and guides the process in a hegemonic form (but not in a dominant form, inasmuch as it depends on Reality-Totality). Conversely, Reality-Totality doesn't dominate the process by *itself*: its self-exposition constitutes itself in the process of knowledge of the subjects, which is an active construction realized by the subjects in Reality-Totality, including the construction of models and theories.

The principle of absolute Speculative Ideality of Reality-Totality and Reality-Totality as living Knowledge

In light of the above, we now introduce the notion of total knowledge of Reality-Totality (the expression of the limit), that we call simply *Knowledge*:

> Principle of Absolute Speculative Ideality of Reality-Totality: *Reality-Totality is identical to Knowledge, or shortly, Reality-Totality = Knowledge.*

If we assume the Principle of Absolute Speculative Ideality of Reality-Totality, we can say that the Knowledge self-exposes itself by an immanent self-organizing form of the subjects. So we have:

> *Reality-Totality is living Knowledge!*
> *And us, we are the self-organizing parts of it!*

As we saw before, we can identify Reality-Totality with the Idea. Therefore, this living and active Knowledge (i.e., Reality-Totality) can also be identified with the Idea and we can denominate the view discussed in this chapter as an *absolute speculative idealism*. Therefore, as discussed previously, the Idea self-exposes itself to us by a self-organizing process of which our proper process of knowledge of it is part.

Notice that this characterization of Reality-Totality as Knowledge implies that it can be (partially) understood, and thus, there will always be reasons, or better, explanations, including those by models and theories, that reveal it. In this sense, it is in consonance with the continuous development of contemporary science and with the possibility of the permanent construction of models.

Final remarks: a possible absolute speculative idealism

We presented here the general philosophical view: *Reality-Totality is living and active Knowledge, and we are (active) parts of the self-manifesting Idea, the Idea which manifests itself in a self-organizing process of which our proper process of knowledge of is part.*

In our view, this conception of Reality-Totality is interpreted as an absolute speculative idealism. As pointed out previously, we admit various forms of interpretation of Reality (and Reality-Totality), in the sense that people can make use of various principles to interpret Reality by certain originative decisions. This leads us to consider our propose here as merely one of various possible interpretations. However, as stated earlier, it is an interpretation that allows for the coordination of all others, since, as noted earlier, for it, all interpretations aim to expose Reality-Totality for themselves, even if they don't admit it.

In closing, we can say that, at its limit, the view presented here leads us to a viewpoint close to Hegel's absolute speculative idealism; the development of this idea, however, is a theme for another work.

Notes

1 For more details about the difference between *scientific knowledge* and *philosophical knowledge*, see Granger (1998); for the notion of *factum*, see Granger (1988), p. 249; on the analysis of the process of *verification*, see Granger (1992).
2 Before elaborating on genetic epistemology, Piaget developed genetic psychology in order to test questions about facts related to epistemology. We maintain here the same spirit of submitting questions of fact to the appropriate sciences.
3 On the notion of *mental imagery*, see Piaget (1964) and Piaget and Inhelder (1966b).
4 The term *operation*, in this chapter, means a mathematical partial function, *i.e.*, a function f that associates to each element x (or list of x elements) of a domain D (in which f is defined) one element y of D; f does not necessarily have to be defined for all elements (or lists of elements) of D.
5 Although *Reality-Totality* and *Idea* are equivalent, we will here use the sign *Reality-Totality*, because this evokes more easily what is signified. We will return to the denomination *Idea* when we characterize our perspective as a speculative absolute idealism.

References

Granger, G-G. (1988). *Pour la Connaissance Philosophique*. Paris: Editions Odile Jacob.
——— (1992). *La Vérification*. Paris: Editions Odile Jacob.
——— (1994). *Formes, Opérations, Objets*. Paris: J. Vrin.
——— (1995). *La Science et les Sciences*. 2^{éme} ed. Paris: P.U.F.

Piaget, J. (1964). *La Formation du Symbole chez l'Enfant: Imitation, Jeu et Rêve; Image et Représentation*. Neuchâtel: Delachaux et Niestlé S.A.

——— (1952). *The Origins of Intelligence in Children*. New York: International University Press. Translation of *La Naissance de l'Intelligence chez l'Enfant*. Neuchâtel: Delachaux et Niestlé S.A.

Piaget, J., and Inhelder, B. (1966a). *La Psychologie de l'Enfant*. Paris: P.U.F.

——— (1966b). *L'Image Mentale chez l'Enfant: Étude sur le Développement des Représentations Imagées*. Paris: P.U.F.

Saussure, F. (1966). *Cours de Linguistique Générale*. Paris: Payot.

PART II

Biophysical and cognitive approaches

6

VITAL FLOW

The Self-Organization stage

Romeu Cardoso Guimarães

Introduction

Living beings and the life process are difficult to define. Both entities are complex, as are the observers of these phenomena. There are many aspects to their components, and their multi-faceted interactions involve them in mutuality. In this chapter, we approach the problem from an evolutionary perspective.

Our origins-of-life model sprang from studies on the formation of the genetic code, specifically, the origins of the association between genes and proteins. These researches focused on the singular ("digital") "letter-by-letter" correspondences (Butterfield et al., 2017) between the triplets of bases in the genetic material which are the codons of messenger mRNAs or the complementary anticodons of the transfer tRNAs, and (Froese et al., 2018) the amino acids that the latter carry with specificity (cognitively) and transfer to a nascent protein chain. The formation of a system of correspondences describes the encoding process. Decoding is accomplished inside cellular ribosomes. This process, called the translation of a sequence of codes into that of proteins, would be better named transliteration, since it involves no interpretation. These correspondences are the first instance of the specificity that characterizes life, allowing the construction of structures and functions via the organization of the sequences. Most studies in code formation take for granted the origins of encoding (Froese et al., 2018), and do not address the question of whether the enzymatic aminoacyl-tRNA synthetase activity was or was not preceded by a ribozyme.

Leading concepts

"*Living beings* are metabolic flow systems that self-construct on the basis of memories and adapt/evolve on the basis of constitutive plasticity. *Life* is the ontogenetic and evolutionary process instantiated by living beings" (Guimarães, 2017). Flow dynamics is a scientific substitute for the old mystical "vital force". Viruses are mobile elements. In Guimarães (2017), there is a technical exposition of the Self-Referential Model (SRM) for the formation of the genetic code; Guimarães and Santos (forthcoming) is a discussion intended for the general reader. These concepts mean that the nucleoprotein system is sustained by metabolism. The system is internal to the cell but is fed from environmental substrates, which indicates that the living is an integral part of geochemical systems. Therefore, the evolutionary flow is universal and includes the biologic or metabolic. This chapter identifies the series of cellular structures and functions that construct the metabolic flow and guarantee its nonstop activity. These serial mechanisms configure a suite of molecular sinks that are also the activities of the living.

Our description ends with the development of cellular reproduction, which is the last component of the sink system. It is also the initiator of the next stage, where Darwinian processes are added to Self-Organization. It is considered that other aspects of living activities and life processes are evolutionary additions to the cellular basics. Most prominent is the development of sexual mechanisms, from meiosis, and of aggregative abilities. These start with multicellularity and open the routes to other elaborations, including the social and psychic. It is suggested that all these aspects should be at the least compliant with, if not promoters of, the flow. Accordingly, the classification of diseases should also benefit from an examination of their impacts on the metabolic flow. The final section of the chapter examines an apparent convergence between models attempting to describe the origins of the three large realms: the quantum, the cosmic, and life.

Construction of the cellular flow system

Proteins are the main cellular components. The system that accomplished their synthesis includes the nucleic acids and is the center of the sink mechanisms. This central sink, the protein synthesis system, has to be maintained as healthy. If not continually or perennially active, it must at least be fully capable of resuming activity as soon as environmental conditions are adequate, in case it has to be temporarily suppressed due to harmful intercurrences.

The first necessity during the period of evolutionary origins was for diversification of protein structures and functions, so as to guarantee energy and amino acid sources in the upstream (nutrition) direction and safe transport of products away from synthesis sites in the downstream direction. There must be no clogging, blockades, or accumulations along the flow routes. The protein synthesis system works as a substrate-stimulated ratchet, and does not function as a drive-forward mechanism in itself. Diversification of proteins depends mainly on gene duplication, genetic mobility, and horizontal gene transfers, and incorporates epigenetic influences. All these mechanisms are grouped under the concept of plasticity, both phenotypic and genomic.

The living mechanism incorporates reversal of the direction of polymerization (that is, degradation via hydrolysis) only for generating monomers at nutritional salvage. Sensitivity to saturation is one of the regulatory processes of polymerization activity. Mechanical saturation is avoided through control of cell volumes and shapes, avoiding the effects of overcrowding through skeletal features such as the microtubules and filaments of eukaryotes, and the cell walls of plants, fungi, and prokaryotes. If saturation does not work by itself, it triggers the activation of repressors. Saturation may not need to be general but may be restricted to some specific kinds of processes that developed the role of critical sensors for control. One of the main metabolic sensors, the mechanistic target of rapamycin (mTOR), is directed precisely to amino acid availability in cellular pools, especially to leucine, which is very abundant in protein compositions.

The flow sectors

The central sink

Vital dynamics are configured as a metabolic flow system. The flow starts at nutrition but is centered on the protein synthesis sink of amino acids and energy, which is kept constantly active and healthy. Nucleic acids, aside from their possibly original role as protein-producing machinery (Figures 6.1 and 6.2), develop the ligation of codes into long polymer strings that work as replicative memories (genes) for protein sequences (Figure 6.3).

The SRM data indicate that elements taken up from the environment in the era of the formation of the code and of the metabolic system were very simple, being of the C1–C2 realm (e.g., methanol, CO_2, acetate) from which more complex internal materials were constructed. The search for the prebiotic equivalents of the present-day compounds that carried the C1 compounds should, thus, focus on the pterin- and folate-like functions.

All kinds of amino acids that would have been formed in abiotic contexts might have participated as substrates or ligands for dimer-directed-protein-synthesis (DDPS) (Figure 6.1), but the quantitative availability of most of them would have been subjected to fluctuations that impeded the construction of codes on their bases. The only firm connection that is supported by the SRM is glycine: it is abundant prebiotically and the first in biosynthesis.

Specificity

Encoding of "letters" is the first instance of biological specificity, which makes possible the construction of genetic sequences that specify structures and functional attributes. Encoding is the result of a long evolutionary development of

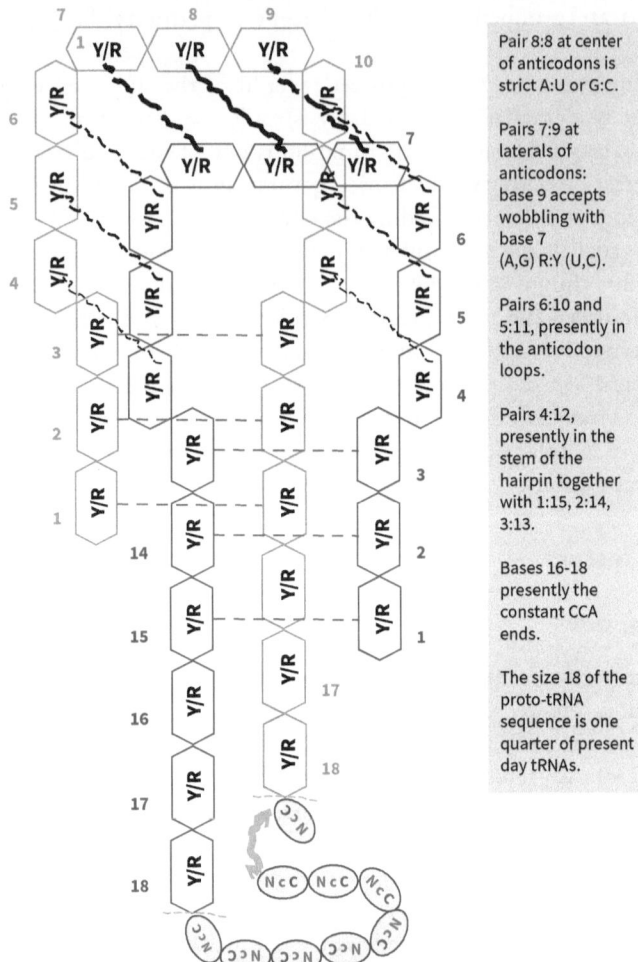

Pair 8:8 at center of anticodons is strict A:U or G:C.

Pairs 7:9 at laterals of anticodons: base 9 accepts wobbling with base 7 (A,G) R:Y (U,C).

Pairs 6:10 and 5:11, presently in the anticodon loops.

Pairs 4:12, presently in the stem of the hairpin together with 1:15, 2:14, 3:13.

Bases 16-18 presently the constant CCA ends.

The size 18 of the proto-tRNA sequence is one quarter of present day tRNAs.

FIGURE 6.1   Dimer-directed-protein-synthesis (DDPS). The proto-tRNAs in the dimer are shown with structure and direction, indicated by the numbering of the bases, to mimic present-day tRNAs' anticodon stem-loop and acceptor stem. Members of the dimers are exchangeable with others in the pool since base pairs are weak and thermally dynamic hydrogen bonds. According to the "singularity" (monomers paired, coherent or superposed) of the state of the pair, there are no definitions in the direction of the transferase reaction, that may be bi-directional, or in the codon *versus* anticodon exchangeable identity. The structure is considered a proto-ribosome: it holds two tRNAs together and facilitates the transferase reaction (double arrow).

an association between a protein – a (proto)synthetase – and its substrate (proto) tRNA (Figure 6.2). The iterative cycles of association reach specificity at some dynamic plateau of the process that is called "cognitive"; the members in the association become cognate to each other. The protein activity is initially (proto)

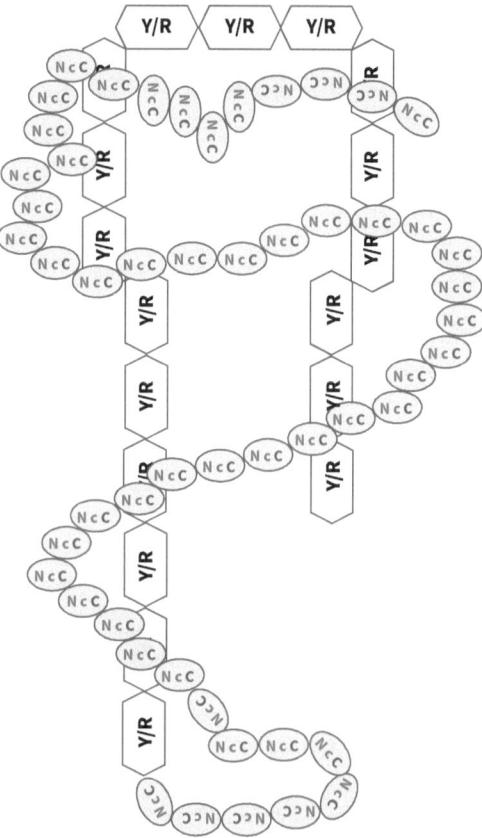

FIGURE 6.2 Decoherence by protein binding and encoding. Association of a product from the DDPS with a proto-tRNA. This is the self-referential aspect in the process and also the mechanism of decoherence produced by proteins. At the ribonucleoprotein (RNP) association, precursor and product bind to each other. The ensemble forms a production system when the protein is stable, and confers stability to the RNP that maintains the protein synthesis activity. The protein has more affinity to one of the proto-tRNAs. The RNP is precursor to a synthetase-tRNA encoding reaction with specificity. Other early associations may be precursors to, e.g., the ribosomal RNPs. A designed viral version of a similar process is in Butterfield et al. (2017).

tRNA binding that evolves into an enzyme that attaches its other substrate, an amino acid, to the (proto)tRNA. In the aminoacyl-tRNA that is formed, the tRNA becomes a carrier of the amino acid that can be transferred to a growing (nascent) peptide or protein. Only after "knowing" how to work with the "letters" (tRNAs and amino acids) could cells start the process of enchaining them into organized sequences that can be decoded (Figure 6.3).

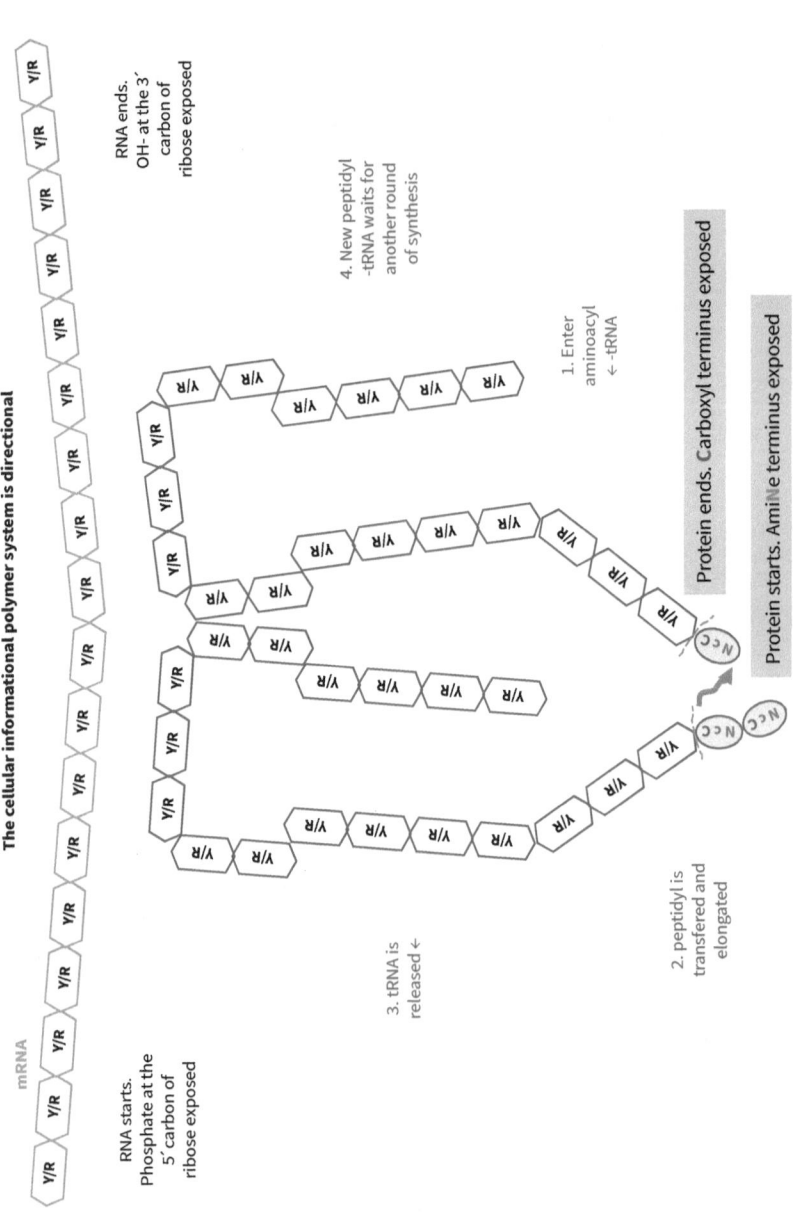

FIGURE 6.3 Decoherence by intromission of mRNA and decoding at ribosomal protein synthesis. The mRNA is a chain of contiguous triplet codes (codons) that are decoded by the anticodons of tRNAs, inside ribosomes. The synthetase reaction and the movement of the ribosome along the mRNA are unidirectional. Evolution of codon contiguity required various torsions and curvatures in the components to reach adequate topological fitting.

A designed version of a similar but viral-like associative process has already been proposed (Butterfield et al., 2017). Our proposal is to start with proto-tRNAs that make proteins, which, in turn, coat the proto-tRNA, thereby building a ribonucleoprotein (RNP) system. This evolves into a cognate functional ensemble. The object envisioned as being at the early state of the process – an RNP globule – may be similar to RNP granules, stress granules, and P-bodies (Hughes et al., 2018; Treeck et al., 2018).

A generic demonstration of specificity is the homochirality in biopolymers. It is required that amino acids in proteins are homogeneous with respect to the "handedness" (hand, in Greek: *keir*) of their structures, in the sense that our hands match one against the other but do not match when superposed. Another analogy is with the movements of clock hands: clockwise means right-handed, counter-clockwise left-handed. The chiral property of amino acids is related to the complexity of the alpha-carbon (the central lower case **c** in the oval amino acid symbol in Figures 6.1 and 6.2). This carbon is simple in glycine, which is non-chiral, but complex in serine, which is left-handed like all other protein amino acids. Conversely, all nucleic acid sugars are right-handed. This property reflects the strict enzymatic requirement for precise and specific 3D-fitting between catalytic pockets and substrate shapes. This homogeneity would be better than mixtures of different 3D structures, possibly guaranteeing speed, smoothness, and repetitiveness in all steps of molecular interactions, thereby being an adequate and necessary participant in the flow dynamics. A useful image is that of the common toboggan-like helical structures of biopolymers; these would be kinky and stepladder-like if built from left-handed and right-handed mixtures.

Protein diversification

A large diversity of cellular structures are directed to guarantee the flow. This starts with an uptake of substrates from the environment and ends with the extrusion of waste into the environment, which is degraded due to both kinds of interference. Environmental modifications are only one among a variety of challenges and stressful conditions that organisms confront from external and internal sources, and which take part in forcing the evolution of the flow system. The cells can only answer with further diversification. Their resources for this reside in the *plasticity* of their components (less extensive in RNA than in proteins, and even less extensive in DNA than in RNA) and of the network organization. The model for the protocell is that of a spongy RNP granule or globule imbibed in water. The internal/external distinction is maintained in the globule through spontaneous protein motility and binding activities.

Crowding without saturation

The aggregating forces among components of the "spongy globule" and in development of surface structures (the membrane function) are rudimentary at the start, and only later guarantee resistance to fragmentation and invasion by water. Under such fragile conditions, the globule cannot grow beyond a certain limit, at which point aggregation and surface tension forces are overcome, resulting in chunks being split from the main body and lost. Otherwise, however, these conditions introduce a stimulatory effect on the flow system, based on reduction of the crowding intensity in the globule.

The process is spontaneous, but functions as though the system itself were avoiding saturation from overcrowding and guaranteeing that the protein synthesis activity keeps a steady pace. This spontaneity, in the case of non-living physicochemical events, is documented by Sydney Fox's microspheres and Alexander Oparin's coacervates. Their gemmulation or budding is similar to the oocyte-polar body or the mother-daughter cell associations of budding yeast. This mechanism of losing chunks of protoplasm, now called the shedding of vesicles or exosomes, became regulated in fully developed cells but has the same stimulatory effect.

Waste

Such spontaneous stimulatory benefit is afterwards combined with the solution of the problem of extrusion of waste so that the chunk-shedding mechanism acquired enriched functionality. Metabolic waste is problematic mostly with respect to nitrogenous compounds. These cannot be transformed into gases and vapor, as happens with hydrocarbon and carbohydrate waste. Nitrogenous derivatives are toxic (ammonia), insoluble (urate), or water-requiring and pollutant (urea). Some amino acids and some proteins are not well reabsorbed by the kidneys and are disposed of.

The problem is greater with some proteins that are most difficult to degrade and cannot be recycled through catabolic processes such as the proteasomal and the lysosomal-autophagic. Degradation intermediates may include indigestible remnants that form entangled aggregates. These may be toxic to cellular organization, especially via the exposition of the unprotected internal hydrophobic protein cores, finally forming amyloid grains and plaques. The solution was their shedding as chunks accommodated in vesicles. These associated benefits are at the origin of the ubiquitous cellular character of shedding vesicles and exosomes.

Exosome multifunctionality includes, through the loss of biomass, the beneficial effect of the protein synthesis sink stimulation, regeneration, and replacement of lost aged material by renovated materials, and the corresponding structures and functions. The stimulatory mechanism is analogous to that obtained from tree pruning.

Extracellular vesicles and exosomes are also seen as communication vehicles that cells utilize for transport of macromolecules inside multicellular bodies, and can be utilized in medical "liquid biopsies" for diagnosis. Intracellular accumulation of protein tangles is seen as a possible causative agent of various diseases, including Alzheimer's and other (mainly neurological) conditions.

The SRM is the first model of the code to consider 3D protein construction pathways – the 3D folding rules – among the tests and components of its structure. Correct folding is important not only for the construction of the native or functional architectures but also for guaranteeing the proper degradation of the proteins without accumulation of toxic intermediates. Empirical data – the N-end Rule – describes which amino acids contribute to protein resistance to degradation, that is, those with the correct folding when placed at their amine-ends. Nascent peptides without the adequate conformation, bearing destabilizing amino acids at the N-ends, are directed to quick degradation. This property shows that the code has a circular structure: initiation and termination codes are the last to form and are dictated one by the other, producing an "informational closure" that is also material.

Reproduction

The cost associated with such losses of protoplasm were partially circumvented when some of the shed chunks received portions of the genetic memories and became daughter cells. This marks the beginning of the evolution of reproductive cycling: losses turned into regulated protoplasmic fission with an inclusion of genomes. Cells acquired the properties of (1) potentially perennial activity of the protein synthesis sink and (2) the installation of the Darwinian process. Reproduction accomplishes various functions: it avoids wasting some of the extruded pieces, such as the vesicles; it guarantees continuity of the individual self-maintenance flow; it is generic-nonspecific, driving the whole individual chain of flow, and installs the population-evolutionary open-ended flow.

Cells that reach large sizes (such as in the G2 stage of the eukaryotic cell cycle) run the risk of having protein synthesis reduced/inhibited due to mass-action or saturation-induced repression. In consequence of the benefits of releasing them from the inhibition and maintenance of activities that contribute to health and productivity, exosome extrusion became regular and obligatory. This is equivalent to the cytoplasmic fission in cell reproduction. This stage was reached when sets of genetic memories – genomes – were added to the chunks of cytoplasm being eliminated, these becoming daughter cells. The original function of the first phase in the reproduction process – cytoplasm fission – is that of regenerating protein synthesis activity, while the second function was that of rescuing the cytoplasm portions from loss by becoming daughter cells.

Bacteria that have had the walls peeled off, L-forms, bear an exposed fragile membrane, and the formation of exosomes is easily observed. In some of these, genome inclusion is found, showing that this is a primitive form of reproduction. Furthermore, it was seen that cell reproduction may be asymmetric with respect to the inclusion in only one of the daughter cells of an "inclusion body", which contains the tangles of damaged and undigested proteins clumped together. This is a simple way of producing healthier lineages, free from the tangles (germ-line analogs) and separated from the less healthy lineages (somatic line analogs).

Behaviors focus on the extremities of the flow system

Understanding cell reproduction as a beneficial by-product of protoplasm loss is another instance of "informational closure". Evolutionary populations are formed when the Darwinian open-ended process is installed; whenever reproduction is active, the protein synthesis activity may be never-ending.

The main environmentally open behaviors, the most evident "vital force" manifestations, are at the extremes of the flow process: nutrition, which feeds the protein synthesis sink, and reproduction, which pulls the sink downstream and keeps it active nonstop. Intermediate mechanisms are internal and organic and may go unnoticed by organisms or external observers, as they are mostly hidden to the organic senses and to the conscious feelings of individuals. The work presented here is dedicated to clarifying these internal and not readily accessible drives.

Realms of the flow

The general idea of the flow is not new, but we add a plausible rational explanation for it, spanning from the entropic universe to the origins of life and to reproduction. The internal drive mechanisms, often not apparent to most observers, are clarified. In medical genetics, the idea of flow is essential to the concept of the inborn errors of metabolism, and we propose that medical science attempt to verify how the flow concept can be applied interestingly to all disease categories and classifications. In the Darwinian account, the reproductive flow is measured as adaptive fitness. We will now pinpoint its centrality to protein synthesis at the cellular-unit level, and generalize the flow concept for all realms of biology.

Cognitive convergence

As previously mentioned, the most salient aspects of the organismal flow dynamics, with stronger appeal to the general observer, are (1) the relational and interactive behaviors, at the openings of the metabolic mechanisms to the environments, mostly at the uptake domains, that is, nutrition and feeding, and (2) the reproductive drive. The psychological counterparts are the obvious ones – desires, impulses, and drives for food and sex – that are consistently accommodated together with the cellular basis.

Such convergence may mean more than just coincidental final results of investigations. The mutual fit indicates that our minds should follow tendencies or biases in favor of repetitions of mechanisms, that is, of the application of the same or similar explanations in a diversity of realms. According to the view presented here, the background to this constancy is engrained in the natural selection mechanisms, which are continually forcing the adjustments and adaptations between organisms/observers and environments/objects of interest. Our minds are biased in this unidirectionality; in the ethological and psychological realms, this would be reflected in certain "cognitive architectures" of minds, configured like some kind of the Jungian archetypes, that is, as modes of apprehension of experiences.

Evidence for these converging routes have arisen repeatedly during our studies, intriguingly enough to raise suspicions of some kind of constraints or directedness/limitations to reasonings or creativity. The initial protein conformations indicated by the SRM were the intrinsically disordered segments. This is consistent with the quantum mechanical rationale that their primal objects – wave packets – are also disordered. In both cases, the order, reflected in informational patterns, would arise at the interaction of entities. The same mechanism shows up in very different realms of study, and different approaches often find a way of fitting together. It seems that our minds can only be relaxed, pleased, and happy, when some kind of "informational closure" is reached; the alternative would allow for sustaining instabilities and loose ends in the lines of reasoning that would create or maintain intellectual tension.

Another closure was reached at the formation of the initiation and termination mechanisms at translation of mRNA. The entire set of elongation codes was formed utilizing a "primitive punctuation system" based on the higher metabolic stability of the protein head segments and the lower stability of the tail segments. The last codes were the specific punctuation: adding one specific anticodon for initiation made the system immediately delete the anticodons that were in conflict with the initiation, whose codon complements became the terminators.

In favor of this convergence, there is also the highly prevalent (and justifiable, within the scientific community) principle of parsimony, which states that multiplicity in the composition of explanations is acceptable only when there is compelling evidence. In situations where evidence is lacking, the principle of simplicity becomes a methodological must. Parsimony reigns, but the propositions based solely on this principle are challenges for the attempts of experimentalists.

Coherence-decoherence

The DDPS process (Figure 6.1) has some peculiarities that are worth being analyzed in themselves and compared with the ribosome- and mRNA-directed processes (Figure 6.3). The (proto)tRNA associations are dynamic, via hydrogen bonding, and may generate different states:

a   States complementary to other (proto)tRNAs, forming the dimers and opening the route to the DDPS (Figure 6.1);

b   States complementary to other RNAs, which may open the route toward translation of mRNA;

c   States with binding to proteins, which inaugurates the RNP associations such as the ribosomal associations and the evolution of the aminoacyl-tRNA synthetases for encoding (Figure 6.2).

The state in Figure 6.1 is called *coherent* or *superposed*, following the terminology of quantum mechanics. The proto-tRNA components are simple, singular, and of the same kind – that is, mutually equivalent – and, therefore, presenting undecided identities and functions: (1) the transferase activity is a-directional or bidirectional, the donor or acceptor functions are interchangeable. Any of the partners may serve the aminoacyl- or peptidyl-carrier functions, and may exchange the functions in each round of the realization of the transferase function, which is a job of the joint pair; (2) the codon and anticodon functions are also interchangeable, coding and decoding being the same. This would have been the only state present at the initial encoding, during the primordia of the formation of early protocell populations.

In the quantum realm, the components of wave packets in a coherent or undecided state may probabilistically produce the classical wave or the particle states (and associated properties) after going through the interactions that lead to decoherence, including those that are part of the detection or measurement processes. States (b) and (c) (see Figures 6.2 and 6.3) are decohered, and each (proto)tRNA may acquire individuality as "classical" components of the cellular translation machinery. The transition from the DDPS to the ribosome- and the mRNA-directed state would involve the intromission of two decohering interactions: one with state (c), the peptide products of the DDPS that may be heterogeneous and able to bind differentially to the oligomers (Figure 6.2); and another with state (b), the entry of another (proto)RNA in the place of one of the members of the dimer, which would be taking the role of the classical mRNA (Figure 6.3).

Three singularities

The explanatory similarity may be extended even more, to reach the third great division of the knowledge of nature, cosmology, which utilizes the same terminology of singularity. Life and quantum mechanics have already been commented upon. All are described by us, the observers – reflexively, in a fourth realm.

In the micro-world, the quantum objects – wave packets – are difficult to describe, almost intangible, and are said to be of undecided (superposed) identity. The interactive events that gave origin to the conversion into the classic wave or particle – with probabilistic distribution – are said to produce decoherence or to detach one from the other component(s) that are no longer superposed. In the macro-world, there would have been a primeval singularity. A very dense and hot object became unstable by itself – of course, there was nothing else with which it could interact – and entered a process of expansion, the Big Bang. Space was extended in between the wavicles, particles and waves. It started a trajectory of progressive cooling, with degradation of different kinds of energy – from the highly dense, e.g., photons, to the less dense, e.g., heat. At some intermediate point in this evolution, living beings appeared.

It is tempting to suppose that the two primeval singularities, the micro and macro cases, would share some characters. There are other names to describe the idea whose basic character is that of some kind of primeval association between distinct states which, submitted to some not well-defined interference, dissociate into our good old classic states. In our cultural traditions, these are the oriental yin-yang complementarity, and the original symmetry plus symmetry-breaking events in physics.

Then enters the third singularity: the proto-tRNA dimer that is supposedly a proto-ribosome and the initiator of the primeval cellular entity. Its dynamics shares various similarities with the singular states of the entities in both of the preceding physical realms. It took us almost a decade to realize how similar the proto-ribosome model was to both of the physical models.

Acknowledgments

I would like to thank the CLE Interdisciplinary Self-Organization Group at UNICAMP. I am especially grateful to Alfredo Pereira Jr., and the kind attentiveness of Jonatas Manzolli and Luiz Oswaldo Carneiro Rodrigues.

References

Butterfield, G. L., Lajoie, M. J., Gustafson, H. H., et al. (2017). Evolution of a designed protein assembly encapsulating its own RNA genome. *Nature*, 552, pp. 415–420.

Froese, T., Campos, J. I., Fujishima, K., Kiga, D., and Virgo, N. (2018). Horizontal transfer of code fragments between protocells can explain the origins of the genetic code without vertical descent. *Scientific Reports*, 8, Article Number: 3532.

Guimarães, R. C. (2017). Self-referential encoding on modules of anticodon pairs – roots of the biological flow system. *Life*, 7(2), Article Number: 16.

Guimarães, R. C., and Santos, F. R. (in preparation). Vital flow dynamics: cell-molecular basics.

Hughes, M. P., Sawaya, M. R., Boyer, D. R., et al. (2018). Atomic structures of low-complexity protein segments reveal kinked β sheets that assemble networks. *Science*, 359, pp. 698–701.

Treeck, B. V., Protter, D. S. W., Matheny, T., et al. (2018). RNA self-assembly contributes to stress granule formation and defining the stress granule transcriptome. *PNAS*, 115, pp. 2734–2739.

7

IMPLICATION AND INFORMATION

A quantitative-informational analysis of material implication

Marcos Antonio Alves and Itala M. Loffredo D'Ottaviano

Introduction

It is common to assign, in semantic terms, the truth values T or F to the formulae of the language of a system. In this chapter, we introduce an informational semantics for the classical propositional logic (CPL) and assign an amount of information to each formula of its language. Our main objective is to show that the usual material implication does not capture the notion of information, as developed in the mathematical theory of communication (MTC) by thinkers such as Shannon and Weaver (1949).

Initially, we present some basic definitions used in our approach. Next, we produce a probabilistic semantics for CPL; we establish a relation between the formulae of a CPL language and the events of a random experiment, from which we define the probability value for each formula of a language; we then introduce the notion of probabilistic implication, examining whether it evades certain paradoxes attributed to the usual notion of implication. Finally, we introduce the notion of amount of information in a formula of a CPL language and verify some results on the implication.

Elements of an axiomatic probability theory

The theory of probabilities, $\mathbb{P}$, deals with random experiments. As a basis for $\mathbb{P}$, we use Zermelo-Fraenkel's usual set theory with the Choice Axiom (ZFC), in which we have elements of the usual number theory (Enderton, 1977).

Definition 1

a   A *random experiment*, denoted by Σ, is one that, when repeated several times, presents different results, denoted by A_Σ, in each repetition.

b   The *sample space* of a random experiment Σ, denoted by U_Σ, is the set of all possible results of Σ.

c   The *number of elements* in the sample space U_Σ, denoted by $n(U_\Sigma)$, is a natural number greater than zero, which indicates the cardinality of U_Σ.

d   A *sample space* U_Σ is *equiprobable* when all its elements have the same chance of occurring.

Definition 2

a   An *event* of Σ is any subset of the sample space U_Σ. The ith event of Σ is denoted by $E_{i\Sigma}$. Thus, $E_\Sigma \subseteq U_\Sigma$.

b   The *number of elements* of an event E_Σ, denoted by $n(E_\Sigma)$, is the quantity of elements of U_Σ belonging to E_Σ.

c   The *event* E_i of Σ, *complementary* to E_Σ, denoted by "$\overline{E}_\Sigma$", is defined by:

$$\overline{E}_\Sigma =_{df} \{A \in U_\Sigma \mid A \notin E_\Sigma\}.$$

d   The *event* E of Σ, *the union of* $E_{i\Sigma}$ *and* $E_{j\Sigma}$, denoted by "$(E_{i\Sigma} \cup E_{j\Sigma})$", is defined by:

$$(E_{i\Sigma} \cup E_{j\Sigma}) =_{df} \{A \in U_\Sigma \mid A \in E_{i\Sigma} \text{ or } A \in E_{j\Sigma}\}.$$

e   The *probability of occurrence* of an event E in the random experiment Σ with an equiprobable sample space U_Σ, denoted by "$p(E_\Sigma)$", is defined by:

$$p(E_\Sigma) = \frac{n(E_\Sigma)}{n(U_\Sigma)}.$$

f   The *conditional probability* of event $E_{i\Sigma}$, given the event $E_{j\Sigma}$, denoted by $p(E_{i\Sigma} \mid E_{j\Sigma})$, is given by

$$p(E_{i\Sigma} \mid E_{j\Sigma}) = \frac{p(E_{i\Sigma} \cap E_{j\Sigma})}{p(E_{j\Sigma})}$$

When $p(E_{j\Sigma}) = 0$, we define that $p(E_{i\Sigma} \mid E_{j\Sigma}) = 0$.

Based on $\mathbb{P}$, in what follows, we develop a probabilistic semantics for CPL, as developed by Shoenfield (1967).

A probabilistic semantics for languages of CPL

We call this perspective, the *probabilistic semantics for CPL* (henceforth, $S_\mathbb{P}$). As shown by Alves and D'Ottaviano (2015), the behavior of $S_\mathbb{P}$ is not strictly equivalent to the behavior of the usual classical truth-functional semantics (henceforth, S_V).

The expression "Form(L)" denotes the set of formulae of a language L; "Var(L)" denotes the set of propositional variables or atomic formulae of L; the letters "φ", "ψ", and "γ" are metalinguistic variables that represent elements of Form(L); "P_0", "P_1", "P_2", etc., are the atomic formulae of L, and "Γ" represents any finite subset of Form(L).

Definition 3

A function f_Σ: Var(L) $\rightarrow \wp(U_\Sigma)$ is a *Σ-situation for L*. The function f_Σ univocally extends to the set of formulae of L, f_Σ: Form(L) $\rightarrow \wp(U_\Sigma)$, as follows:

a   If φ is atomic, then $f_\Sigma(\varphi) = E_\Sigma$, defined by f itself;

b   If φ is of the form $\neg\psi$, then $f_\Sigma(\varphi)=\overline{f_\Sigma(\psi)}$;

c   If φ is of the form $\psi \vee \gamma$, then. $f_\Sigma(\varphi)=f_\Sigma(\psi)\cup f_\Sigma(\gamma)$

Each formula of L is associated with a single event in a given situation f_Σ. However, distinct formulae can be, and in general are, associated with the same event in f_Σ.

Definition 4

The *probability function of a formula* φ according to f_Σ, denoted by "$P_\Sigma(\varphi)$", is given by the probability of the event E in the sample space Σ associated to φ by f_Σ, according to the definition of probability function p in $\mathbb{P}$:

a   If φ is atomic, then $P_\Sigma(\varphi) = p(E_\Sigma)$, such that E_Σ is $f_\Sigma(\varphi)$;

b   If φ is of the form ψ, $P_\Sigma(\varphi)=p\left(f_\Sigma(\psi)\right)$;

c   If φ is of the form $\psi \vee \gamma$, $P_\Sigma(\varphi) = p(f_\Sigma(\psi) \cup f_\Sigma(\gamma))$.

Example 1

Consider the random experiments Σ_1 and Σ_2, such that $U_{\Sigma 1} = \{1, 2, 3, 4, 5, 6\}$ and $U_{\Sigma 2} = \{C, K\}$. Then:

TABLE 7.1 Probability value of formulae

φ	$f_{\Sigma 1}(\varphi)$	$P_{\Sigma 1}(\varphi)$	$f_{\Sigma 2}(\varphi)$	$P_{\Sigma 2}(\varphi)$
A_1	$\{2, 4, 6\}$	½	$\{C\}$	½
A_2	$\{1, 3, 5\}$	½	$\{K\}$	½
A_3	$\{1, 2, 3, 5\}$	⅔	$\varnothing$	0
A_4	$\{1\}$	⅙	$\varnothing$	0
A_5	$\varnothing$	0	$\varnothing$	0
$A_1 \rightarrow A_2$	$\{1, 3, 5\}$	½	$\{K\}$	½
$A_1 \rightarrow A_3$	$\{1, 2, 3, 5\}$	⅔	$\{K\}$	½
$A_2 \rightarrow A_3$	U	1	$\{C\}$	½
$A_5 \rightarrow A_4$	U	1	U	1

Theorem 1: For every f, we have (for the proof, see Alves, 2012):

a $P(\overline{\varphi}) = 1 - P(\varphi)$

b $P(\varphi \rightarrow \psi) = p\left(\overline{f(\varphi)} \cup f(\psi)\right).$

c $P(\varphi \rightarrow \psi) = p\left(\overline{f(\varphi)}\right) + p\left(f(\varphi)\right) \times p\left(f(\psi) \mid f(\varphi)\right).$

d $P(\varphi \rightarrow \psi) = 1$ if, and only if, $f(\varphi) \subseteq f(\psi)$.

e   If $P(\varphi \rightarrow \psi) = 1$, then $P(\varphi) \leq P(\psi)$.

From Theorem 1.c, we can conclude that, in most cases, $P(\varphi \rightarrow \psi) \neq p(f(\psi) \mid f(\varphi))$. Exceptions occur, for example, when $P(\varphi) = 1$ or $f(\varphi) \subseteq f(\psi)$.

Next, we develop an analysis of the notion of implication, especially from the probabilistic semantic perspective. We draw a parallel between the results from this perspective and those from the usual semantic perspective, S_V, based on the notion of truth value. We also deal with two interpretations of material implication, based on S_V and $S_{\mathbb{P}}$. We call the interpretation derived from S_V *veritative-functional material implication,* and the interpretation derived from $S_{\mathbb{P}}$ *probabilistic material implication.*

In both $S_{\mathbb{P}}$ and S_V, the material implication, $\varphi \rightarrow \psi$, has a maximum value (1 in the case of probability value and T in the case of truth value) in a circumstance (situation or valuation, as usually defined in the literature) if, and only if, φ has a minimum value (0 in the case of probability and F in the case of truth) or ψ has a maximum value in the given circumstance. In the opposite sense, the value of $\varphi \rightarrow \psi$ is minimal if, and only if, the value of φ is maximum and the value of ψ is minimal. In addition, the values of $\varphi \rightarrow \psi$, $\varphi \vee \psi$, and $\neg(\varphi \wedge \neg\psi)$ must be equal to each other under any circumstances. These characteristics can be proved for S_V from the CPL definitions. In S_P, they can be easily shown, where it also immediately follows that $P(\varphi \rightarrow \psi) = P(\neg\varphi \vee \psi) = P(\neg (\varphi \wedge \neg\psi))$.

Despite certain correspondences, $S_{\mathbb{P}}$ and S_V have substantial semantic differences. The first and most evident concerns the nature of the value assigned to the formulae. While in S_V the value is a truth-value, "T" or "F", in $S_{\mathbb{P}}$ this value is a number on the rational continuum [0,1]. Even if it is possible to draw a parallel between truth and probability, they are distinct concepts with different meanings. In terms of implication, this change of perspective produces specific results for $S_{\mathbb{P}}$, as the last three results of Theorem 1 illustrate.

Theorem 1.d is specific to $S_{\mathbb{P}}$. This result has no corresponding result in S_V, because in S_V, the truth value of a formula in a given valuation is defined directly by the truth value of its constituents. There is no recourse to an associated element, as in the case of $S_{\mathbb{P}}$, where the probability value of a formula depends on the probability value of the events associated with its constituents and cannot be found directly from their probability value.

Given that if $P(\varphi) \leq P(\psi)$, then $P(\varphi \rightarrow \psi) = 1$ is not valid, the following case is also invalid: if $P(\varphi \rightarrow \psi) \neq 1$, then $P(\varphi) \nleq P(\psi)$. In S_V, the corresponding result holds: if $V(\varphi \rightarrow \psi) = F$, then $V(\varphi) > V(\psi)$. This case also is a result in $S_\mathbb{P}$: if $P(\varphi \rightarrow \psi) = 0$, then $P(\varphi) > P(\psi)$, given that under these conditions $P(\varphi) = 1$ and $P(\psi) = 0$. The two perspectives have the same behavior for this case when we deal only with extreme values, 0 or 1.

The notion of a reasonable formula is defined from its probabilistic value: the closer the probability value of a formula is to the maximum, the more reasonable it is. Next, we show that there is not always symmetry between the reasonableness of a probabilistic implication and the content relation of its constituents.

Theorem 2 (for the proof, see Alves, 2012):

a   If $P(\varphi) \leq P(\gamma)$, then $P(\varphi \rightarrow \psi) \geq P(\gamma \rightarrow \psi)$.
b   If $P(\varphi) \leq P(\gamma)$, then $P(\psi \rightarrow \varphi) \leq P(\psi \rightarrow \gamma)$.

Whenever $f(\varphi) \subseteq f(\psi)$ and $f(\gamma) \subseteq f(\psi)$, then $P(\varphi \rightarrow \psi) = P(\gamma \rightarrow \psi)$, independently of the value of φ and γ. Except for this case, Theorem 2.a ensures that the less reasonable the antecedent of a probabilistic material implication, the more reasonable is the probabilistic material implication itself. This reasonableness, however, is determined solely by the probability of the antecedent and generally does not guarantee the relation of content between the constituents of the material implication, as exemplified later.

Considering the situation of the throw of dice, for example, one can say that the sentence "if it shows the number one, then it shows an even number" is more reasonable than the sentence "if it shows an odd number, then it shows an even", although, in terms of content relation, both look the same. It may also occur that in a given situation the statement "if it shows even, then Érico Veríssimo is a writer" is more reasonable than the statement "if it shows a prime number, then Érico Veríssimo is a writer", although none of them have any relation of content between the antecedent and the consequent. Finally, the statement "if the moon is made of cheese, then Érico Veríssimo is a writer" is totally reasonable, that is, has probability one, when evaluated in a situation where the probability of the sentence "the moon is made of cheese" is zero.

The second item in Theorem 2 shows that what was said earlier about the reasonableness of the antecedent of a material implication, and about the reasonableness of the implication itself, can also be attributed to its consequent. It explains that the more reasonable the consequence of a material probabilistic implication, the more reasonable is the probabilistic material implication itself, as we illustrate later.

By considering the situation of the throw of dice again, the sentence "if it shows an even number, then it shows a number greater than two" is more reasonable than "if it shows an even number, then it shows a number less than or equal to two", although, in terms of content relation, both look the same. It may also occur that in a given situation, the statement "if Érico Veríssimo is a writer, then it shows a prime number" is more reasonable than the sentence "if Érico

Veríssimo is a writer, then it shows an even number", although none of them have any relation of content between the antecedent and the consequent. Finally, "if Érico Veríssimo is a writer, then it rains or it does not rain" is totally reasonable, that is, it has probability one, because the probability value of "raining or not raining" is maximum.

Sometimes the reasonableness of a formula ensures the relation of content between its constituents. This is the case of the material implication between two atomic formulas, $A_i \rightarrow A_j$, when $P(A_i \rightarrow A_j) = 1$, $P(A_i) \neq 0$ and $p(A_j) \neq 1$. Here, first, by Theorem 1.d, we have that $f(A_i) \subseteq f(A_j)$; second, the event corresponding to the antecedent is possible in the given situation; third, the event corresponding to the consequent is not necessary in this situation. What the antecedent of the implication says, the consequent also says.

As in S_V, in S_P the material implication also presents paradoxical results (in the sense of what goes beyond opinion, that is, of what does not coincide with common sense or which contradicts intuition) when compared to the intuitive notion of conditional. To emphasize this, we introduce a different conception of material implication and will attempt to ascertain its scope and limits. To deal with this different type of implication, we need to extend the L language in order to express this concept through a new connective in L. This extended language, denoted by "$L^|$", is itself L plus the symbol "$|$" in its alphabet.

The definitions of $L^|$ are the definitions of L plus the clauses referring to the symbol $|$. In the usual definition of L-formula (CPL), we add the following clause:

e   if ψ and φ are formulae of $L^|$, then $\psi | \varphi$ is formula of $L^|$, called *probabilistic implication* of ψ by φ.

The formula $\psi | \varphi$ must be read as φ implies probabilistically ψ; φ is called the antecedent and ψ the consequent of the implication.

To Definition 2, we add the following clause, referring to the *probability value* of a probabilistic implication formula, denoted by "$P_\Sigma(\psi | \varphi)$":

$$\text{Definition 2': } P_\Sigma(\psi \mid \varphi) = p(f_\Sigma(\psi) \mid f_\Sigma(\varphi)).$$

When the probability value of the probabilistic formulae is determinate, we investigate how much φ interferes in ψ. It means assessing to what extent the occurrence of $f(\varphi)$ interferes with the occurrence of $f(\psi)$.

Example 2

Based on Example 1, we have:

In Example 2, considering the usual definition of number, it can be inferred that the occurrence of the event as an even number excludes the possibility of showing an odd number. Thus, the probability of showing even number

TABLE 7.2 Probability value of implications

If φ, then ψ	$P(\varphi \rightarrow \psi)$	$P(\psi \mid \varphi)$	Translation
If A_1, then A_2	½	0	If it shows an even number, then it shows an odd number.
If A_1, then A_3	⅔	⅓	If it shows an even number, then it shows a prime number.
If A_2, then A_3	1	1	If it shows an odd number, then it shows a prime number.
If A_5, then A_4	1	0	If it shows the number seven, then it shows the number one.

implying the showing of an odd number should be zero. That is, the probability value of the first implication of this example would have to be zero, as with the probabilistic conditional, but not with the probabilistic material conditional.

The second formula of Example 2 also illustrates a discrepancy between the probabilistic material implication and the intuitive notion of implication. We intuitively perceive that the appearance of an even number considerably restricts the appearance of a prime number. However, in this case, unlike the probabilistic conditional, the value of the probabilistic material conditional is quite high.

In the third formula, the probability value of the two conditionals is equal, since in this specific case, the "odd number" event is contained in the "prime number" event. If $f(\varphi) \subseteq f(\psi)$, then $P(\varphi \rightarrow \psi) = P(\psi \mid \varphi) = 1$. Here, the value of both conditionals captures the intuitive notion of implication.

Finally, the last formula again illustrates a paradoxical result regarding probabilistic material implication: a formula with a minimum value implies any formula. In the case of probabilistic implication, the value is minimal, since, according to this implication, empty implies nothing. From our perspective, this seems intuitively appropriate. The probability of something happening, given the occurrence of an event whose probability of occurrence is zero, is effectively zero. Another possibility would be to consider such conditional probability to be indeterminate in these cases; we prefer not to assume this possibility.

Another paradoxical result of the probabilistic material implication, when compared to the intuitive notion of implication, is the following: again considering the non-biased throw of dice, the statements "if it shows number one, then it shows an even number" and "if it shows an odd number, then it shows an even number" have different probability values according to probabilistic material interpretation. However, intuitively, in both statements there seems to be no implication between the antecedent and the consequent. Thus, the probability value of each should be zero, as occurs when these formulae are interpreted as probabilistic implications.

Information and material implication

In this section, we define the amount of information in a formula, and present some results on material implication.

Definition 5

The *amount of information*, or *informational value*, of a formula φ of L according to a situation f_Σ, denoted by "$I_\Sigma(\varphi)$", is the numerical value defined by $I_\Sigma(\varphi) = -\log_2 P_\Sigma(\varphi)$. When $P_\Sigma(\varphi) = 0$, we define that $I_\Sigma(\varphi) = 0$.

Example 3

As in the previous example, consider the random experiments Σ_1 and Σ_2. Then:

TABLE 7.3 Amount of information of formulae

φ	$f_{\Sigma 1}(\varphi)$	$I_{\Sigma 1}\,\varphi$	$f_{\Sigma 2}(\varphi)$	$I_{\Sigma 2}\,\varphi$
A1	{2,4,6}	1	{H}	1
A2	{1,3,5}	1	{T}	1
A3	{1,2,3,5}	0,58	$\varnothing$	0
A4	{1}	2,58	$\varnothing$	0
A1 → A2	{1,3,5}	1	{T}	1
A1 → A3	{1,2,3,5}	0,58	{T}	1
A2 → A3	U	0	{H}	1

The informational value of the first probabilistic material implication above is the unit, because the probability of the event corresponding to the formula in both situations, which is the same event of its consequent, is ½. In both situations, by Definition 5, we have that $P(A_1 \to A_2) = P(A_2) = $ ½ if, and only if, $I(A_1 \to A_2) = I(A_2) = 1$. However, from the intuitive point of view, what is being said in the sentences "if it shows an even number, then it shows an odd" and "if it shows heads, then shows tails" seems impossible to happen. Thus, the probability value of that implication should be zero and, consequently, its amount of information should also be zero. This paradox arises from the definition of probabilistic material implication, more specifically from the probability value defined for these formulae, as we have already discussed.

The informational value of the second implication according to $f_{\Sigma 1}$ is less than the value of the previous formula in this same situation, because the informational value of the consequent is smaller. This formula, in the given situation, would be translated as "if it shows an even number, then it shows a prime number". As discussed earlier, the decrease in value is not due to the relation between the parts of this material implication, but it occurs simply because of the change in the value of the consequent.

The statement in the previous paragraph shows that the second material implication does not capture the notion of the amount of information suggested in the MTC. Intuitively, this sentence says something quite difficult to happen: the occurrence of a prime number after the occurrence of an even number. Therefore, the information in this sentence should be quite high according to the Shannonian perspective, contrary to what happens in the case in question.

Finally, the last material implication, interpreted in the light of $f_{\Sigma 1}$, says that "if it shows an odd number, then it shows a prime number". As this is obvious in this case, it is correct to understand that the amount of information should be null. Already, according to $f_{\Sigma 1}$, it says that "if it shows tails, then it shows the number one" or "if it shows tails, then the moon is made of cheese". The amount of information in this statement depends solely on the amount of information in the antecedent. This case shows that the amount of information of a material implication is independent of the content relation of its parts.

It can be easily shown that $I(\varphi \rightarrow \psi) = I(\neg \varphi \lor \psi)$. However, in quantitative terms of information, it would not always be appropriate to identify the informational value of the two formulae in question. Considering the situation of a non-biased throw of dice, the amount of information in the statement "if it shows a prime number, then it shows an even number", for example, should not be equal to that of the statement "it does not show a prime number or it shows an even number". In fact, the amount of information of the implicative statement, in this situation, should be much greater than the information present in the disjunctive statement. The occurrence of an even number in the hypothesis of the occurrence of a prime number is extremely unlikely; however, the event associated with the statement "it does not show a prime number or it shows an even number" is much more likely to occur.

The previous result shows that by making the sentences "$\varphi \rightarrow \psi$" and "$\neg \varphi \lor \psi$" equivalent, CPL does not capture the notion of information as defined in MTC. Contrary to the equation in the previous paragraph, the reduction of uncertainty in these two formulae may be different. The amount of information in them may be different in certain situations.

In addition to the problem of the possibility of sentences like "$\varphi \rightarrow \psi$" and "$\neg \varphi \lor \psi$" having different amounts of information, they may also differ in their information content (roughly, in what they mean). When uttering: "if there is extraterrestrial life, then I am a triptych", for example, a speaker may not be saying "there is no extraterrestrial life or I am a triptych". The implication, in this case, categorically means "there is no extraterrestrial life". In defining material implication through disjunction and negation, or the inverse, defining disjunction through material implication and negation, CPL does not capture either the quantitative or qualitative aspect of the information in these types of formulas. The next example is another case that illustrates this situation.

Example 4

As in the previous examples, consider the random experiment Σ_1. Then:

TABLE 7.4 Probability and informational values of implications

If φ, then ψ	$P(\varphi \to \psi)$	$I(\varphi \to \psi)$	$P(\psi\,|\,\varphi)$	$I(\psi\,|\,\varphi)$	Translation
If A_1, then A_2	½	1	0	0	If it shows on evens, then it shows on odds.
If A_1, then A_3	⅔	0,58	⅓	1,58	If it shows on evens, then it shows on primes.
If A_2, then A_3	1	0	1	0	If it shows on odds, then it shows on primes.
If A_5, then A_4	1	0	0	0	If it shows number seven, then it shows number one.

In a throw of dice, the events showing even and showing odd are mutually exclusive. Therefore, when it shows an even number, it is impossible for an odd number to occur. That is, "shows even" never implies "shows odd". The first implication in Example 4, therefore, says something impossible. According to the notion of information adopted in MTC, the amount of information in this formula should be zero, as with probabilistic implication.

The second probabilistic material implication in Example 4 has a smaller amount of information than the first, solely because the amount of information of its consequent is less than the amount of information of the consequent of the previous implication. According to this situation, the probability of showing an even number implying the showing of a prime number is very low, but not impossible, as in the previous sentence. The second sentence is very bold and says something quite risky, which eliminates many possibilities. Thus, its probability should be low and hence its quantity of information is rather high. This is what happens with probabilistic implication, but not with probabilistic material implication.

In the cases here considered, the occurrence of the event associated with the antecedent of the implication does not alter the subsequent sample space. When calculating the value of a probabilistic implication, $\psi\,|\,\varphi$, we assume the $E_{j\Sigma}$ event associated with the antecedent of the formula as if it were the sample space of a new random experiment, which we shall call $U_{\Sigma'}$. The $E_{i\Sigma}$ event associated with the consequence of the implication is seen as if it were transformed into an event $E_{i\Sigma'}$, consisting of the elements of E_i that are also elements of $E_{j\Sigma'}$, that is, $E_{i\Sigma'} = E_{i\Sigma} \cap E_{j\Sigma}$. Then $p(E_{i\Sigma}|E_{j\Sigma}) = n(E_i(\Sigma'))/n(U(\Sigma')) = P(\psi\,|\,\varphi)$. In the second formula of the example, we evaluate, within the quantity of even numbers, how many are prime, which in this case is in the ratio of one to three and where the result of the probability value of the formula is ⅓. In the probabilistic material implication, there is no such relation of dependence between the antecedent and the consequent of the formula. This is one of the reasons why its informational value seems not to be adequate.

The third implication in Example 4 states something obvious. Since every odd number is prime, in this specific situation, it is absolutely certain that odd means prime. Thus, in this specific case, both implications capture the information notion of MTC.

Finally, in the last sentence in Example 4, the two versions of implication have the same amount of information, but for different reasons. In probabilistic material implication, this sentence is necessary; in the version of probabilistic implication, it says something impossible. Since, in both cases, there is no reduction of uncertainty, the amount of information of the two sentences is zero. The difference in probability value of both was discussed earlier.

Next, we make a probabilistic and informational comparison of the probabilistic material implication and probabilistic implication in which some of its constituents are probabilistically valid or contradictory.

Example 5

According to the Table 7.5, although the significance of each of the implications is distinct, in most cases the probability value and the informational value of the probabilistic material implication and the probabilistic implication are the same. The shaded columns in Table 7.5 indicate the differences in values between the formulae. In the first shaded region, while the probabilistic material implication is necessary, the probabilistic implication is impossible. In probabilistic terms, these two formulae are opposite, but in informational terms, they have the same value; the former because it says something obvious, and the latter because it says something impossible. From the point of view of probabilistic material implication, a contradiction implies anything. From the point of view of probabilistic implication, a contradiction implies nothing. It can be observed that, for all Σ, I_Σ $(\varphi | \bot_\mathbb{P}) = I_\Sigma (\neg(\bot_\mathbb{P} \rightarrow \varphi))$.

In the second shaded region, we observe a difference in the probability value and the informational value of the two implications. As in the veritative-functional material implication, the value of a probabilistic material formula of the form $\varphi \rightarrow \bot_\mathbb{P}$ depends on the value of the negation of φ. In this case, the more reasonable is φ, the less reasonable is the implication. Thus, the less

TABLE 7.5 Comparison between probability and informational values of formulae

ψ	$\bot_\mathbb{P} \rightarrow \varphi$	$\varphi | \bot_\mathbb{P}$	$\top_\mathbb{P} \rightarrow \varphi$	$\varphi | \top_\mathbb{P}$	$\varphi \rightarrow \bot_\mathbb{P}$	$\bot_\mathbb{P} | \varphi$	$\varphi \rightarrow \top_\mathbb{P}$	$\top_\mathbb{P} | \varphi$
$P(\psi)$	1	0	$P(\varphi)$	$P(\varphi)$	$P(\neg\varphi)$	0	1	1
$I(\psi)$	0	0	$I(\varphi)$	$I(\varphi)$	$I(\neg\varphi)$	0	0	0

informative the formula φ, the more informative the probabilistic material implication formula $\varphi \rightarrow \perp_{\mathbb{P}}$. In the case of the corresponding probabilistic implicational formula, since it is impossible for a formula to imply probabilistically a contradiction, the amount of information in this formula is null.

Final considerations

The main results obtained from the elements presented in this work are:

a   Material implication and probabilistic implication are not equivalent and may have different results in the same situation;

b   Material implication, when treated in terms of $S_{\mathbb{P}}$, does not always have the same results as when treated in terms of S_V. There are some results that are specific to $S_{\mathbb{P}}$, given their inability to be stated in terms of S_V; and that there are some results shared by the two semantic perspectives. Furthermore, some results are characteristic of one perspective or another, due to the nature of the value (probability or truth) attributed to the formulae. When we deal only with extreme values (0 or 1, corresponding to truth values F and T), $S_{\mathbb{P}}$ and S_V share relations relative to the value of a formula and the value of its constituents.

c   In S_V, there is the possibility of the absence of the relation of content between the elements of a sentence, even when it comes to valid formulae, that is, one whose value is always true. We show that this characteristic also remains in S_P, paying particular attention to probabilistic material implication.

d   We show that in a large number of cases there is no relation between the reasonableness of an implication and the connection of content between its constituents. This same result can be proved for conjunction, negation, and biimplication. The material implication, from the probabilistic point of view, is also subject to the paradoxes of material implication, when evaluated in S_V.

e   We show that probabilistic material implication does not capture the intuitive notion of implication when it is understood as a classical causality relation. This notion presupposes some kind of connection between the cause (antecedent) and the effect (consequent). However, it can be said that, just as with veritative-functional material implication, the implication does not express, nor does it intend to, the notion of causality.

f   We show that the material implication does not capture the notion of information as suggested in the MTC. Certain sentences have a very different informational value than they should present. One of the main reasons for this is the lack of a dependence relation between antecedent and consequent in material implication. As a result, CPL itself does not capture such a notion of information. A logical system with a probabilistic implication can perhaps do so more appropriately.

References

Alves, M. A. (2012). *Lógica e Informação*: uma análise da consequência lógica a partir de uma perspectiva quantitativa da informação. PhD thesis, Universidade Estadual de Campinas, Campinas/SP.

Alves, M. A., and D'Ottaviano, I. M. L. (2015). A quantitative-informational approach to logical consequence. In: Koslow, A. and Buchsbaum (eds.) *The Road to Universal Logic: Festschrift for the 50th Birthday of Jean-Yves Béziau, Studies in Universal Logic series* (pp. 105–124). Switzerland: Birkhäuser, Springer International Publishing.

Enderton, H. B. (1977). *Elements of Set Theory*. San Diego, CA: Academic Press.

Shannon, C. E., and Weaver, W. (1949). *The Mathematical Theory of Information*. Urbana: University of Illinois Press.

Shoenfield, J. (1967). *Mathematical Logic*. Reading, MA: Addison Wesley Publishing Company.

8

SELF-ORGANIZATION IN COMPUTATIONAL SYSTEMS

Ricardo Ribeiro Gudwin

Introduction

The overall definition of a *system* can be quite inclusive. According to Bresciani Filho and D'Ottaviano (Chapter 3, this volume), a system can be any unitary entity of a complex and organized nature, constituted by a non-empty set of active elements, which maintain some relation to each other and have time-invariant characteristics that give them, as a whole, an identity; in other words, a set of elements forming a structure with a given functionality, which gains an identity due to providing this functionality.

Systems can be *natural*, as in the case of living systems, or *artificial*, as in the case of technological artifacts built by men. The *organization* of a system pertains to the structure connecting the many different system parts, meaning that each part is in some kind of relation to one or more other parts of the system. This connection might be physical, or simply logical, implying that there is some relation between two specific parts within the system. The term *organization* can be etymologically traced back to the process leading to the formation of *organs* in a living body, i.e., a collection of parts of a physical body collectively exhibiting some sort of functionality, and gaining together the status of an organ due to performing this functionality (the Greek word *organon* means: *tool*). In a man-made technical system, the idea of *organization* might not be really related to true *organs*, as in a living body, but have to do with the existence of *sub-systems* of an overall system, carrying on some sort of functionality and having in themselves some sort of identity as a composite of parts.

These definitions are definitively quite abstract but can be applied to either living beings, technical artifacts, or social entities like business organizations or groups of people. But it is important to point out that the term *organization* holds a double meaning. It might designate either a fixed structure defining the system (an atemporal structure or, in a temporal system, its particular structure in a given instance of time), which we will be calling here an s-organization (meaning an organization as a structure), or it can be related to a dynamical process, where this structure is modified through time, which we will be calling here a p-organization (meaning an organization as a process). In other words, the term *organization* might be related either to the structure configurating a system in a given instant of time (as, e.g., when we talk about the *current* organization of a system, an s-organization), or to the process where this structure might be changing over time (as, e.g., when we talk about the process of the organization of a business company department throughout its history, a p-organization). This double meaning can be quite misleading, especially if we are talking about a specific phenomenon: Self-Organization.

According to Debrun (Chapters 1 and 2, this volume), a system can be *hetero-organized* if the system structure is imposed by an external source, or *self-organized* if it produces itself, i.e., if it evolves by itself as an offspring of an interaction of a set of parts which were independent before they gained together the status of an entity (its identity as a system) due to providing a functionality. Debrun also differentiates between *primary Self-Organization*, where the system gains its identity *during* the initial interaction of its many parts, before they can be seen as a system, and *secondary Self-Organization*, where the system, after its identity is already determined, is able to preserve its structure if this structure is disturbed in any way.

Using our terminology, Self-Organization is achieved if a system might reach a stable s-organization due to its p-organization, or, similarly, if it reaches its s-organization due to its p-organization. If this stable s-organization is achieved during the p-organization process, primary Self-Organization is achieved. If the system is originally hetero-organized, but is able to maintain its stable s-organization due to its p-organization, when its structure is disturbed in some way, we might say that the system reaches only secondary Self-Organization.

Usually, machines and other man-made systems are hetero-organized, as their structure doesn't change over time. This is the most common case for man-made systems. Conversely, natural systems like living beings can be seen as (primary) self-organized systems. It is important to point out, though, that a system does not necessarily have to have material parts. The parts can be just logical ones. An interesting case we want to investigate here is the case of computational systems; not just any kind of computational systems, but specifically *software*. Our first guess regarding this kind of system is to necessarily classify them as hetero-organized. This is because software usually originates from the work of a human programmer, which externally imposes on the software a fixed structure which does not change, gives it its identity as a system, and provides some sort of functionality. But we would like to investigate whether some sort of

Self-Organization is possible in software systems. And, if this Self-Organization is possible, would it be just secondary Self-Organization, or might primary Self-Organization also be possible?[1]

Some characteristics of software systems

As we have already pointed out, software systems are intrinsically different from other kinds of systems, because of their lack of materiality. A software system is just pieces of data stored in a computer memory, being executed by a computer processor. These data might be stored in files, or be already loaded in the computer's memory. Even though there is some sort of materiality, because these data are stored in a physical memory, this is not what gives a software system its identity. If software is moved to a different kind of memory, it will still be the same software; software is, therefore, an example of purely logical systems.

Let us start our analysis with some characteristics, which are common to all software systems. Even though a piece of software is just data on the memory, in fact, it is important to split this data into two different categories which are functionally distinct. In any software system, part of the data is what we call *code*, i.e., instructions in processor language (machine language) which are to be executed by the processor during the software execution. The rest of the data is *data proper*, i.e., data contents which are meant to be used as variables in the software execution. They might contain some initial data, which could be important for software execution, or be just a placeholder for the data, which will be processed during software execution. Now, from a systemic point of view, we might define that *code* embeds the structure of the system, and *data proper* is just a set of states which describes the system's inner working. In a traditional hetero-organized software system, the code is generated by a programmer, conceived usually in a high-level programming language and compiled into bytecodes in order to constitute the system structure. This structure never changes during system execution, and the system provides its functionality, as conceived by design.

But things might become a little fuzzy if we start considering a special class of software systems: self-modifying systems. Because computer instructions can change the contents of memory in general, and the code is stored into memory, it is possible to build software programs, which are able to modify their own code. Usually, software systems are not supposed to do this, but it is feasible, and, in fact, it is done in some cases, which we will be detailing later. If we consider that the code is the system structure, this opens up the possibility for the system to change its own structure, and then further opens up the possibility of having self-organized software systems.

We would also like to point out some characteristics typical of two different kinds of software: *virtualizers* and *simulators*. *Virtualizers* are software programs, which are used as functional equivalents of a whole computer running a different kind of system. Virtualizers are becoming very popular as a way of testing multiple kinds of systems within the same host computer. In a so-called *host system*, the virtualizer creates a virtual machine, functionally indistinguishable from a real machine (except for performance issues), where another system, called the *client system*, can be installed and executed. Using a virtualizer, it is possible, for example, to run a Linux system (as a client system) on top of a host system running MS-Windows. The user might think she is running a Linux machine, but, in fact, the real computer is running MS-Windows. Aside from issues related to computer performance (because a virtualized system runs slower while compared to its host system), the user can have a completely Linux experience, even though in the background there is really an MS-Windows system being executed. Virtualizers are important in our case, as they help us to understand that it is possible to run a completely different system as a client system virtualized on top of a host system. This idea is important for us because we can conceive a host system, which is hetero-organized, virtualizing a client system that might be a self-organized system.

Simulators are somewhat similar to virtualizers, but with a distinction. In the case of a simulator, many properties of the real system being simulated are the same, both in the real system and its simulation, but some of these properties are not. For example, a simulator can preview the amount of rain in a given location, over time, but the simulated rain is not wet and does not have some other properties of real rain. Simulators are very important in engineering, because they are very effective in allowing predictions of real system behavior, without the burden of having available a real system which might be expensive or time-consuming. Simulators are important for us because even though there might be properties of the real system which are not present in the simulated system, if we build the simulation with care, the most important properties (those we are interested in studying) can be brought into the simulation, and in this case, the simulated system embeds in itself everything which is important to us for deriving conclusions. In other words, if we are building a simulation of a real process, which seems to be self-organized, there is a chance that the simulated system is also self-organized.

A final concern must be addressed as well. A computer is inherently a deterministic machine, which means that given the same initial conditions, a computer program will always generate exactly the same behavior. Apparently (Bonabeau et al., 1997), randomness is a requirement for self-organized systems.[2] If we want a computational system to be self-organized, how can we conciliate these facts? The issue of randomness has, for a long time, been a concern in the construction of simulators (James, 1995; Wang, 1996). The solution to this problem is in the use of pseudo-random number generators (Hull and Dobell, 1962), which from a given seed as input, typically exhibit statistical randomness while being generated by an entirely deterministic causal process. Even though the sequence is deterministic, if we use a random seed as, for example, the millisecond in which the simulation was started by the user, the sequence will be completely different each time the program using it is run. From a statistical point of view, the sequence provides a behavior, which can be classified as random.

Object-oriented adaptive software

Object-orientation is an important metaphor used for the construction of software systems. In this metaphor, the software can be seen as a set of objects, interacting among themselves by means of passing messages. The user provides events, like clicking the keyboard or moving the mouse, and these events are sent to interface objects in the system, creating a cascade of messages which will result in the system functionality. We can conceive an object-oriented program as a virtualized system on top of an operational system. The reaction to messages can be either deterministic, in some cases, or random, in other cases, using pseudo-random generators. Objects might have their internal variables changed, and either create new objects or destroy other objects during their behavior. In principle, object-oriented systems are supposed to be hetero-organized, because they need to be developed by a computer programmer. But supposing a process of virtualization, with the aid of pseudo-random number generators, it is possible to conceive that a self-organized client system can evolve on top of a hetero-organized host system. Instead of working with a pre-designed set of objects, we might think of some sort of adaptive object-oriented system, where new objects can be created and the possible interaction between objects is governed by random number generators. In the same way that in a physical system a set of physical rules governs the possibilities of the interaction of physical objects, thus paving the way for the evolution of self-organized systems, a given set of rules can be programmed in object-oriented systems. Furthermore, in this case, with the aid of randomness brought by pseudo-random number generators, self-organized object-oriented systems can be conceived, and with a bonus: in material systems, the rules are those that exist, but in an object-oriented adaptive software systems these rules can be changed and tested, letting different rules govern the dynamics among software objects. Hetero-organized rules work just like physical laws in the material world. The resulting system, however, can be self-organized.

Both primary Self-Organization and secondary Self-Organization are possible. In the case of secondary Self-Organization, an initial structure (an initial configuration of objects) is hetero-organized, and the objects are put to interact with each other. After that, this initial structure might evolve over time, changing its structure but maintaining its identity. But even primary Self-Organization is possible. We might have a completely random generation of objects, and random possibilities of interaction between these objects, as is common in a special kind of systems called evolutionary systems. We provide a better understanding by means of some examples in the next sections.

Learning software systems

Let's start our investigation of this possibility of finding Self-Organization in software systems with a particular kind of software: Learning Systems. Learning

Systems started to appear in the context of Artificial Intelligence, and nowadays are becoming very popular, gaining their own field of research: Machine Learning. A typical learning system is what we call a *neural network*. A neural network comprises a network of entities called artificial neurons, which are abstractions of real neurons found in the brains of living creatures. In some sense, they are a kind of a simulation of a network of real neurons. An artificial neuron does not have all the properties of a real neuron, but they share with them many properties, which might turn an artificial neural network into a self-organized system.

A neural network typically exhibits only secondary Self-Organization. This happens because the initial structure of a neural network is usually hetero-organized, and after the neural network starts operating it changes this structure in order to adapt and learn. But there are specific neural networks, called constructive neural networks, which usually start with just one neuron. In networks of this type, new neurons are created and incorporated to the network as it interacts with its environment. Such constructive neural networks appear to have all the conditions for being classified as primary self-organized systems.

Evolutionary systems

Another kind of system worth mentioning here is the class of *Evolutionary Systems*. Evolutionary systems are a kind of intelligent systems where many aspects of biological evolution are simulated in a computational environment. Despite its many variations, evolutionary systems usually comprise a population of elements, which are processed in an iterative way, using combinatory operators like crossover, mutation, and others, generating new populations over time. Each element in the population represents the solution of a mathematical problem, in the form of a structured collection of parameters, which are important for characterizing the problem. This element is usually called a chromosome, in a direct analogy to biological evolution, which can be evaluated by a *fitness function* that provides a measurement for the quality of the solution brought by a particular element. With this method, many different possible solutions can be evaluated for a given population, and usually the best solution is considered as an output of the evolutionary system. The initial population is usually randomly generated, and as soon as new populations are generated through this evolutionary process, solutions of a better quality are derived (or "evolved"), making the evolutionary system a kind of optimization process. Each evolutionary step implies a first step where the population first grows by means of the many combinatory processes (such as crossover, mutation, etc.), and later contracts when a selection process maintains in the final population only those elements considered more apt, i.e., those with a greater fitness value. With this expansion/contraction movement, the size of a population can be maintained constant throughout the generation of successive populations.

Evolutionary systems, as in the case of neural networks, can be classified as self-organized systems. We can identify both primary and secondary aspects of

Self-Organization in their functioning. As populations are initially randomly generated (using, of course, pseudo-random number generators in their implementation in computers), they may require many steps of evolution before reaching their final configuration, where a good solution for a mathematical problem is generated. This process characterizes the system's primary Self-Organization properties. After that, once good solutions are available in the population, the evolutionary system preserves those of a better quality, thus also meeting the criteria for being classified as secondarily self-organized.

Discussion

We have presented some examples of computational systems where apparently some sort of Self-Organization process seems to be operational. As we pointed out, the first impression would be that true self-organized processes might be impossible in computational systems. This impression comes from the fact that computational systems are programs stored in a computer memory; if these programs are generated by a human programmer, they have a fixed structure and could not be classified as self-organized systems. More than this, these programs lack some kind of materiality, in contrast with the more common kinds of self-organized systems, such as biological systems, where materiality is usual. A third argument against the possibility of self-organized systems in computational systems is the lack of true randomness, as computers are deterministic machines, where a system with the same set of inputs will always exhibit the same kind of behavior as output.

However, we should remember that the definition of a *system* implies that a system is just an abstraction for a fragment of reality. This opens up the perspective of a system being an abstraction for some generic sort of support (a helper for our understanding of reality), this support being a fragment of reality, or another system (i.e., another abstraction). This new perspective allows us to understand the notion of virtualization among systems, where multiple layers of virtualization are possible until we ground them on a fragment of reality. Having these multiple layers in mind allows us now to speculate on the possibility of having self-organized systems virtualized on top of hetero-organized systems. And this is the key for us to understand how self-organized systems could be realized in computational systems. Even though the host system is hetero-organized, this host system embeds another client system, which is virtualized or simulated, and this virtualized/simulated system can be self-organized.

There is still the problem of true randomness. In this case, however, it is necessary to recall that with the use of pseudo-random number generators, and with a true random seed, e.g., the precise millisecond when the program is started, a pseudo-random number generator can generate a unique sequence of numbers which are in principle indistinguishable from a truly random sequence. This leads us to ask two questions. First, does randomness really exist in the natural world or is it just sequences with the same property as those generated by a pseudo-random number generator? Second, are we really sure that true randomness is a strong requirement for Self-Organization, or are the properties given by pseudo-random number generators sufficient for Self-Organization to appear?

Apparently, some kinds of software systems, such as object-oriented adaptive software, learning systems, and evolutionary systems, in being considered as virtualized/simulated systems on top of some sort of hetero-organized support system, might fulfill all the requirements for being considered self-organized. They might have a (pseudo) random s-organization (i.e., a random initial structure) and a set of laws governing their behavior provided by the hetero-organized support system, which is equivalent to the physical laws governing a material system. This s-organization has means for changing itself over time, implementing a p-organization that might conduce this s-organization to the formation of stable units, giving the software system identity through providing it with functionality. This p-organization can also maintain the stability of this s-organization if the latter is disturbed by some external inputs, giving the system properties required for secondary Self-Organization. In summary, such software systems apparently hold all the requirements for being classified as self-organized, either primary in some cases, or just secondary in other cases. The hetero-organization provided by human programmers of this software can be compared to the fixedness of natural laws imposed on material entities in a biological self-organized system, i.e., they can be considered as just "given", with the difference that in the natural world, these laws are unique and cannot be changed, while in the computational world, we are able to explore alternative realities, something which is not possible in the case of the material world.

Conclusions

In conclusion, we propose that some kinds of computational systems, e.g., adaptive object-oriented systems, learning systems, and evolutionary systems, hold all the requirements for being classified as self-organized systems. The argumentation provided in this text does not completely prove this is the case, but brings strong evidence that these systems fulfill the requirements for being classified as self-organized.

It is important to point out, though, that further investigations are still necessary. First, are pseudo-random number generators indeed enough for Self-Organization to appear? Or might there be some property of such pseudo-random sequences that is absent and preventing true Self-Organization to evolve? It will also be important to point out what this property is, and why it is preventing true Self-Organization from appearing, if that be the case. Second, we must explicitly identify what the specific conditions of these systems are that enable them to become self-organized.

Hopefully, with the continuation of this investigation, we will be able to have a better comprehension of what Self-Organization is, and what the minimum requirements for a system are in order for it to be classified as self-organized.

Notes

1 It is important to mention here that even though we are referring to Debrun's notions of primary and secondary Self-Organization, we will also be exploring new possibilities which were not imagined at the time of his original proposals. Some authors might disagree that primary Self-Organization is possible in the cases discussed in the present work.
2 Even though Debrun (Chapter 1, this volume) didn't required true randomness, but only *chance*, in the sense of Cournot. *Chance* in this context means the existence of statistical independence among the interacting parts, allowing new forms (not previously determined by a general law) to emerge.

References

Bonabeau, E., Theraulaz, G., Deneubourg, J. L., Aron, S., and Camazine, S. (1997). Self-Organization in social insects. *Trends in Ecology & Evolution*, 12(5), pp. 188–193.

Hull, T. E., and Dobell, A. R. (1962). Random number generators. *SIAM Review*, 4(3), pp. 230–254.

James, F. (1995). Chaos and randomness. *Chaos, Solitons & Fractals*, 6, pp. 221–226.

Wang, Y. (1996). *Randomness and Complexity*. PhD thesis, University of Heidelberg.

PART III
Practical issues

9

SELF-ORGANIZATION AND ACTION

A systemic approach to common action

Mariana Claudia Broens

Introduction

Ever since the era of classic modernity, the mechanistic perspective and the analytical procedure, its methodological correlate, have been widely adopted in natural sciences and in philosophy. In the realm of the natural sciences, mechanicism conceives the cosmos as a machine whose working is governed by rigorous and precise laws, in contrast to the vitalistic and teleological approaches of the modern period. From the mechanist perspective, the universe is investigated as if it were composed of causally interconnected parts whose workings can be explained by the laws of physics. In the realm of philosophy, the mechanistic perspective, as opposed to the theocentric dogmatism of earlier times, was advocated by modern Western thinkers as an instrument for understanding nature. Many of these thinkers, such as Galileo, Hobbes, Descartes, and Leibniz, went beyond philosophical speculation and produced works in various branches of scientific knowledge, particularly in physics, mathematics, and anatomy.

The mechanistic approach also supposes that in order to investigate and propose explanatory models of the universe, the most adequate approach is to adopt the analytic procedure. This procedure allows one to disassemble the organic or inorganic "machines", decomposing them into their parts in order to understand their causal interactions. The second rule of Descartes' method, "to divide each of the difficulties I would examine into as many parts as possible and as was required in order better to resolve them" (Descartes, 1998, p. 11), was recognizably inspired by mathematics and is a prime example of this procedure.

Classic mechanism, however, is further accompanied by an ancient underlying ontological thesis, metaphysical dualism, the doctrine according to which being is conceived as constituted by two kinds of entities, the material and the immaterial. This idea dates back at least to Plato and can be stated as follows:

a   There are no *qualitative* differences between organic and inorganic entities, for all of them are constituted by the same material elements; all entities vary only in the proportion of their material elements and their degree of organizational complexity.
b   Matter by itself is not conscious, independently of its degree of organizational complexity.

Ontological dualism stems from the presupposition that matter, even that which composes a living being, would constitute much too rough a substratum to be able to serve as the basis for cognitive processes. Matter being considered essentially as "non-thinking" from this point of view, the only remaining alternative for the explanation of cognitive processes was that they belonged to a plane of "non-material" reality. (Recall in this regard that only after the nineteenth century did chemical and physico-chemical studies begin to foresee the high degree of complexity proper to the organization of matter.) Furthermore, the thinkers who advocated these ideas generally conceived of themselves as being the only bearers of conscious reason and moral consciousness, and supposed their condition to be distinct from the state of nature in which other animal species were believed to exist. As a result, the so-called spiritual life in its cultural and social manifestations was conceived as being indissociable from (and almost identical to) human moral life and its underlying exercise of the will.

Thus, it was supposed that entities of two distinct natures, material and immaterial, submitted to different laws: the laws of classical mechanics applied to material beings, and moral laws applied to immaterial ones. Mechanical laws were, therefore, thought to refer to the movements of physical bodies, in general, and to the forces that act on them, while moral laws concerned voluntary actions and moral choices. The prime examples of immaterial entities were reason and the soul (or human spirit), which were believed to be responsible for the production of judgments, free will, and morally valuable action, in sum, for the production of knowledge and culture.

The mechanistic approach was thus born from an ontological schism that put body and mind, nature and culture, physics and metaphysics, on different planes of reality, thereby generating some of the main problems investigated by philosophy since then. Metaphysical dualism, however, was not a mere gratuitous proposal or the simple result of dogmatic assumptions on the nature of being. It was, in fact, considered a reasonable, although not unproblematic, way of avoiding infinite regressions and of giving closure, even if in a precarious way, to the explanatory models proposed by natural science and philosophy. As Gilbert Ryle (1949) observes, philosophers like Descartes ended up building a para-mechanical model to describe the properties of immaterial entities, and this model was similar to the mechanical model proposed for understanding the nature of physical entities. The properties of immaterial entities were, therefore, conceived by denying the properties of the material ones: if material entities occupy a place in space, immaterial ones do not; if material entities are subject to physical laws, immaterial ones are not; if material entities are accessible to public observation, immaterial ones are not; and so on. Both worlds, however, have in common the kind of interaction among the entities that inhabit them: both material and immaterial entities are governed by causal "forces" that allow, according to the dualistic doctrine, interaction among the mechanical and the para-mechanical worlds; and this interaction is supposedly felt by the subject, as Descartes (unsuccessfully) tried to explain in the *Sixth Meditation* (Descartes, 1998).

Thus, despite the theoretical efforts of its proponents, one of the main difficulties faced by metaphysical dualism (even in its more moderate versions) concerns exactly the problem of explaining causal interactions between the material and the immaterial planes of reality. The entities on these different planes obey different laws (the mechanical and the moral) and operate causally within their own ambits (the mechanical and the para-mechanical).

From the end of the nineteenth century onward, the philosophical-scientific scenario changed, insofar as natural sciences began to provide theoretical tools that allowed both for a renewal of existent explanatory models and for the proposal of new ones. Such models are framed in an ontological context denominated physicalism, according to which, as Kim observes: "everything that exists in the spacetime world is a physical thing, and that every property of a physical thing is either a physical property or a property that is related in some intimate way to its physical nature" (Kim, 1999, p. 645). In this context, physicalist mechanicism tries to break free from its dualistic ontological bonds and seeks to restore unity to our conception of being. This project can be considered praiseworthy from many points of view, mainly because it appeared to dissolve the problems posed by metaphysical dualism with regard to the causal interaction between distinct types of substance. It is not without problems, however, as we aim to show in the next section.

Physicalism

One of the main problems with physicalism concerns one of its direct consequences, the widening of the scope of mechanicism. More precisely, mechanistic approaches expanded to cover the plane of action. From this widening of scope it follows that in the dualistic ontology, as we have seen, actions considered voluntary have their own set of regulating laws, that is, moral laws. In the new perspective, however, they are conceived, and consequently start to be investigated, under the aegis of mechanical laws.

At first, the amplification of the scope of mechanicism seemed promising, for it pretended to overcome the 2,000-year failure of moral philosophy to define the laws of practical rationality. Numerous philosophers have proposed systems which attempt to define moral laws without assuming dogmatic or relativistic features, but the sheer amount of attempts (stretching at least from Plato's *Meno* to Kant's in *Groundwork of the Metaphysics of Morals*) seems to be a good indication of the problems faced in characterizing morally valuable action and its regulatory principles in the context of metaphysical dualism. Despite the profound differences among the doctrines of those philosophers, there seems to be a common element among them, that of considering action as the *effect* of will. We can say, roughly, that dualistic moral philosophies conceive the will as a faculty of the soul, which possesses a causal force in relation to the body and is responsible for the execution of actions; the idealization and planning of actions, however, belongs to the soul.

Gilbert Ryle (1949) argues against such a dualistic conception of the will, aiming to show that it is based on a categorial mistake, that is, on a logical misunderstanding that classifies something as belonging to a category to which it effectively does not belong. Thus, the dualistic approach that conceives will as a faculty or as an attribute of a thinking substance not subjected to physical laws, but which has a causal role regarding action, is not sustainable in Ryle's understanding because there would not be a thinking substance belonging to the same category as the extended substance.

From the perspective of mechanistic physicalism, the difficulties faced by moral philosophy are the consequence of a primary misunderstanding: there are no moral laws that order human will, only mechanical laws that regulate the movements and interactions of bodies. It is not strange, therefore, that philosophers had hopelessly tried to find norms that guide human action with a lawlike capacity analogous to that of the mechanical laws. From the physicalist perspective, this project has a problem that condemns it to failure.

The physicalist approach seems to supply satisfactory answers to the gaps in the dualistic moral philosophies. One example is the discovery that neurological lesions can affect patterns of moral behavior, a phenomenon that Michel de Montaigne had already referred to in his *Essays* at the end of the sixteenth century. As Damasio (2006) points out, the clinical case of Phineas Gage showed that changes in the moral behavior of an agent could stem from neurophysiological damages, which implies admitting that patterns of action do not belong to a plane of immaterial reality distinct from the mechanical universe. From this perspective, it is clear that scientific researchers should investigate the neurological mechanisms responsible for our moral life.

Furthermore, the broadened mechanistic approach of physicalism had another significant repercussion, especially in the cognitive sciences. According to Dupuy (2000), the scientific revolution brought about by the cybernetic movement and artificial intelligence, which began in the 1940s, has taken the human species out of the center of research on cognitive processes, in general (formerly set in the context of an anthropocentric dualistic ontology), by assuming that artificial systems could instantiate such processes. Thus, one of the underlying theses of the dualism of substances to which we previously referred, the idea that matter by itself is not conscious, seems to have been definitely surpassed by the cognitivist branch of physicalism.

The great project of the comprehension of action by physicalist mechanicism is that of investigating the nature of action in accordance with the basic thesis of that approach, which, as stated earlier, is that "everything that exists in the spacetime world is a physical thing, and that every property of a physical thing is either a physical property or a property that is related in some intimate way to its physical nature" (Kim, 1999, p. 645). This scientific and philosophical enterprise, however, has revealed itself to be extremely difficult, because, for example, it seems counterintuitive to admit that actions (generally said to be deliberate or intentional) and reflex actions (seen as purely mechanical) could be explained by means of identical regulatory principles.

Contemporary approaches to the philosophy of action that investigate this and many other difficulties of a non-dualistic conception of action can stem from different philosophical postulates. However, as Juarrero (2002) observes, they all seem to have some common elements: (1) an action is the effect of a previous interaction which serves as its cause; (2) an intentional action always causes a branched network of events such as: raises the arm, touches the wall, looks for the light switch, and turns the light on; (3) in this network of events, there may occur actions that are not properly intentional, but are part of the intentional network or are related to the original intention of the agent, such as the (not properly deliberate) action of touching the wall in order to find the light switch.

In this general scenario, different theories of action aim to handle a series of issues involving these intentional causal networks. The condition for an agent to be considered responsible for his or her actions would be, in this general sense, that such actions incurred from a deliberation free of constraints, that is, of a *willing* that occurred without duress. However, if this causal account of action seems to provide a reasonable solution to moral problems, we face numerous dilemmas and paradoxes when we begin to question, for example, what this deliberation consists in and how we can recognize a truly voluntary action (relevant questions from the philosophical point of view, but perhaps even more relevant from the point of view of social and legal interaction).

A famous example of such paradoxes was created by Chisholm (1966), in order to show some of the problems related to this conception of action as part of a causal sequence. This is the example of Carl, who wishes to kill his uncle in order to inherit his fortune. Carl believes that his uncle is at home and drives there, but gets so agitated by the perspective of committing a deliberate homicide that he drives dangerously. On the way, he hits and kills a pedestrian without intending to, and the deceased pedestrian turns out to be his uncle. In this case, there is a question: would this action be a part of a causal sequence related to a premeditated homicide or to a sequence related to an involuntary homicide? The difficulty (almost an *aporia*) is that good arguments can be posited in defense of both alternatives.

Juarrero (2002) observes the causal approach to action is inserted by its proponents, frequently by default, into a context that is ontologically compromised by mechanistic presuppositions that are incapable of satisfactorily explaining the difference between intentional actions and reflex actions.

We understand, furthermore, that the difficulties pointed out here stem from a second problem, a remnant of classic mechanicism that the physicalist approach did not reexamine when it supplanted the dualistic approach. This concerns one of the central theses of mechanicism, which postulated that there are no *qualitative* differences between organic and the inorganic entities, because all of them are constituted by the same material elements, varying only in proportion and complexity of organization.

In contrast, we will argue here that in order to understand action, as well as to understand a significant part of cognitive processes in general, we need to consider that the (evolutionarily constituted) organization of living beings and the degree of complexity inherent to them generates *qualitative differences* that distinguish them from inorganic entities. In order to elaborate our position, we will seek help from evolutionary biology and systemics.

Biology

As in other natural sciences, in the case of biology, the use of the mechanistic approach is long-standing. As Souza (2007) observes, ever since the seventeenth and eighteenth centuries, "organisms could be described as the sum of their components, like machines, or better, 'biological machines', different from the mechanisms built by human beings only in reason of a difference in materials". From the twentieth century on, with mechanicism and the analytical procedure continuing as a background, biology has directed its investigations to the elements that compose organisms (cells, proteins, genes) and to the functions of these elements.

In recent years, the mechanistic approach has especially flourished in the area of genetics, in consonance with studies in the areas of artificial intelligence and cognitive science. As Kelso and Haken (1995, p. 157) state:

> For many geneticists and biologists, the teleonomic character of the organism is due specifically to a *genetic program*. [...] all that we need to know is that a program exists that is causally responsible for the goal-directedness of living things.

According to this perspective on genetics, there is an algorithmic determination on the micro level of the genes, the causal action of which resonates on the macro level of behavior. To its advocates, this perspective justifies a reductionist approach to the phenomena of life and action, in the sense that it is on the micro level that the key to explaining these phenomena is to be found, thanks to the allegedly successful mechanicist approach to action.

Thus, the well-known analogy between the artificial and the natural machine (which, as we have seen, was already postulated in the seventeenth century) reappears in a certain sense as a functionalist mechanism allied to genetics: just as a program supplies a machine with a set of instructions that determine its working, the *genetic program* likewise operates with a similar causal power in relation to organisms.

As observed in the previous section, if in the natural and the cognitive sciences the mechanistic approach brought undeniable contributions to our understanding of a whole set of physical and biological phenomena, in other fields of knowledge mechanism was not as successful and shows signs of exhaustion. As an example of the limits of the mechanistic approach, we could cite the investigation of phenomena whose degree of complexity involves relational properties and qualities that escape the mechanistic approach's explanatory power. Such seems to be the case with many biological, cognitive, social, and cultural phenomena. An example of these complex phenomena, among others, are ecosystems, whose study involves not only so-called *biotic* factors (animals, plants, microorganisms) but also their relations with *abiotic* factors (the soil, the wind, the climate in general); another example is the aforementioned difference between reflex actions and intentional actions (Juarrero, 2002).

Thus, according to our brief outline, if we propose to understand the nature of the actions of organisms (and a whole set of related notions, such as autonomy, intentionality, and responsibility), we cannot do it by presupposing theses and procedures compromised by dualistic ontology, nor can we do it on the basis of mechanistic physicalism. In this context, our approach will seek to investigate organism/environment interaction from a systemic perspective, according to which the action of organisms constitutes a second-order biological phenomenon resulting from self-organized processes.

As we pointed out, given the need to deal with phenomena regarding which the mechanistic approach seems of little help, the investigative tools provided by the systemic approach and the theory of Self-Organization have come to be used increasingly. To some extent, the situation described by Kelso and Haken in the 1990s, when processes of Self-Organization in open systems received little attention from biologists, has gradually changed. We could say the same about philosophers, especially with regard to those who study the nature of mental processes from an interdisciplinary perspective.

By adopting the systemic approach, we believe that we can overcome at least some of the stalemates that, as we have shown, face the approaches to action proposed by ontological dualism and by mechanism. In order to do so, we will aim to show that the systemic approach to action, within the perspective of evolutionary biology, reintegrates action with the dynamics of pattern constitution that is proper to it, in contrast to the analytic-mechanistic framework that considers it part of a branched causal network. The self-organized process of pattern constitution will assume, then, special relevance for the comprehension of action.

Following Ashby (1962) and Debrun (Chapters 1 and 2, this volume), we understand that certain categories of phenomena involve self-organized processes, that is, they arise as result of the spontaneous establishment of a new organization or system through the dynamics of interaction among their constituent elements, without the interference of a central controller or organizer. Debrun observes that:

> There is Self-Organization every time that, from the encounter of actually (and not analytically) distinct elements, a certain unsupervised interaction (or one without an omnipotent supervisor) occurs, and when that interaction eventually results in the constitution of a "form" or in the restructuring by "complexification" of an already existing form.
>
> *(Debrun, Chapter 1, this volume)*

For Debrun, self-organized systems establish a new form of organization not previously determined by their initial conditions. Such systems present three distinct steps: first, the establishment of the system implies a rupture in relation to previous conditions, because the elements that constitute it start to interact, thus creating organization; the second step occurs when the system achieves some kind of stability and, in the words of Debrun, is endogenized; and the third step is the crystallization of the system, which happens when it builds an identity of its own.

According to Debrun, there are two different kinds of Self-Organization: the primary and the secondary. For Debrun, "Primary Self-Organization occurs when the interaction followed by casual integration happens among totally distinct elements (or, at least, among predominantly distinct elements), in a process without subject, or central element, or immanent objective" Debrun, Chapter 1, this volume).

The most relevant aspects of primary Self-Organization refer to its spontaneous appearance and its essentially dynamic nature: a self-organized system is constituted as a result of the interactions of its elements without answering to dictates by controllers or supervisors. The process of secondary Self-Organization, in turn, happens when the system acquires stability and is directly associated with the potential that the system develops for dealing with novelty. In this sense, Debrun observes that:

> Secondary Self-Organization occurs when, in a learning process (corporal, intellectual, existential, or other), the interaction occurs between the parts ("mental parts" and/or "corporal parts") of an organism – and the distinction between parts is, thus, "semi-real" – under the hegemonic, but not dominative, guidance of this organism's "subject-face".
>
> *(Debrun, Chapter 1, this volume)*

We have pointed out that processes of Self-Organization are characterized by the appearance of a form or system by reason of the interaction among the elements without the action of an almighty controller. Learning is established by a change of behavioral patterns resulting from the incorporation of a skill. But we cannot forget that both processes are interdependent: in order for primary Self-Organization to occur, it is necessary that "really distinct" (Debrun, Chapter 1, this volume), and not only analytically distinct, elements begin to interact spontaneously. When such interactions stabilize, the emerging system will be able to instantiate secondary Self-Organization processes.

When an organism is secondarily self-organized, distinct or semi-distinct elements of the system interact for the preservation of the dynamic balance that serves to maintain the system's identity. This balance involves complex forms of interdependent adjustments that preserve the delicate balance between the maintenance of the existing form and the possible assimilation of novelties: both the assimilation of novelties in excess, and the lack of novelties, can bring the system to a collapse.

When it comes to organisms, the self-organized process of the establishment of patterns of action occurs through the countless environmental interactions that seek to preserve the system, the most fundamental among them being interactions that involve metabolic processes and the maintenance of homeostatic equilibrium. An example that illustrates the nature of these self-organized interactions is that of breathing in aerobic organisms: with variations in time, and according to the specificities of each organism, all organisms must fulfill a basic need for oxygen (we do not breathe as a result of the will to breathe, but as a result of the self-organized action of the evolutionarily constituted physiological structure of the body).

In this sense, Ashby (1962), referring to the notion of Self-Organization and to the distinction between "good" and "bad" organization, observes that organizations are not intrinsically good or bad: even the evolutionary development of specialized organs such as brains, intestines, or blood vessels, cannot be considered intrinsically good (as many biologists, according to Ashby, have had the tendency to think). Such specializations were only possible, Ashby observes, because Earth is a planet with an atmosphere; otherwise, according to this writer:

> we would be incessantly bombarded by tiny meteorites, any one of which, passing through our chest, might strike a large blood vessel and kill us. Under such conditions a better form for survival would be the slime mold, which specializes in being able to flow through a tangle of large twigs without loss of function. Thus the development of organs is not good unconditionally, but is a specialization to world free from flying particles.
>
> *(Ashby, 1962, p. 265)*

Organizational potentialities are relative to the specific ecological situation in which organisms develop and in which they find the elements necessary for the maintenance of their equilibrium. To Ashby, every environment is potentially adequate for the apparition of intelligent organizations if it provides stable conditions, but an organization is relative to the environment that fostered its appearance.

In investigations of the constitution of patterns of interaction among organisms and of the role of their individual actions in the constitution of collective patterns, the approach proposed by the theory of Self-Organization has been adopted more and more frequently (Maini and Othmer, 2000; Visscher, 2003, among others). One of the reasons for the growing use of Self-Organization as a theoretical reference point in biology and philosophy of biology is its power to explain dynamic phenomena, among which organism/environment coevolution is one of the main examples.

In this sense, Visscher (2003) observes, "Self-organized systems can evolve by small parameter shifts that produce large changes in outcome" (p. 799). Visscher highlights that, especially in the organizational systems of some insects (bees, for example), the patterns of organism/organism and organism/environment interaction result in an evolutionarily forged Self-Organization process that maintains the system's adaptive and functional effectiveness.

As von Foerster (1960), Atlan (1979), and Schrödinger (1944) have observed in their characterization of the dynamics of organic systems, external disturbances, noise, and/or disorder play a fundamental role in the evolution of living beings, because they incentivize the organization and functionality of systems, just as order, according to Schrödinger (1944/1997), allows for their stability. From the perspective of these authors, the emergence or updating of patterns of action (i.e., individual and collective habits) would stem, on the individual level, from the progressive development of embodied skills as the result of learning; on the level of the species, such emergence or updating of patterns of action would stem from the processes of adaptation and natural selection.

While environmental dynamics poses new challenges to the survival of organisms on the ontogenetic level, the long-term plasticity provided by self-organizing processes on the phylogenetic level gradually shapes their patterns of action. The adaptive success of emergent patterns stems from the constant adjustments performed by the organism/environment system, which are continually on the verge of criticality, in a far from equilibrium state (Kelso and Haken, 1995). This power of adjustment gives them the flexibility necessary to anticipate possibilities of action, and not only of reaction. The authors observe that:

> behavior itself emerges from coordinated actions that promote survival of the individual and hence the species. Here and elsewhere it has been shown that certain basic forms of coordination are subject to principles of self-organization. Might, then, the genotype-phenotype relation eventually be construed in terms of shared, self-organized dynamics acting on different timescales?
>
> *(Kelso and Haken, 1995, p. 157)*

Moreover, the authors observe that biological structures are multifunctional: "the same set of components may self-organize for different functions or different components may self-organize for the same function" (Kelso and Haken, 1995, p. 140). In the understanding of the authors, the multifunctionality of organic systems significantly increases the potentiality of organisms for performing adjustments, increases their possibilities of action and, concurrently, their complexity. An example of this increasing complexification is seen in the development of capabilities related to organisms' aptitudes for collective interaction.

The systemic approach to patterns of action outlined here has the virtue of placing such patterns in a theoretical context, which is different from that proposed by mechanism but inserted in a non-reductionist physicalist perspective that seeks to take into account the *qualitative* dimension proper to organisms. This systemic/physicalist perspective allows for the possibility of understanding certain phenomena of life as emergent from organic substrates and forms of organization alone, and as consequently bringing into being, as noted earlier, properties of the system that are irreducible to the elements that constitute it.

Some of the problems to which we referred previously may be overcome, or at least more clearly stated, by no longer considering deliberate or intentional action as isolated, or even as belonging to a branched network. There seems to be in action a qualitative element related to temporal factors proper to the biological history of organisms that escapes the explanatory range of mechanism.

We understand that it is quite possible that the notion of deliberation inherent to action itself is intimately related to the search for the satisfaction of basic organic needs that characterizes the environmental interaction of living beings. In particular, actions related to the search for food, along with the consequent metabolic processes related to nutrients, can possibly be considered as the first step in the long evolutionary history of the intentional actions of organisms. This primordial search for nutrients may be the root of a kind of proto-intentional generator of patterns of action shared by organisms.

Our proposal that the recognition of organisms' patterns of action have their evolutionary origin in this pattern of food search may allow us to deepen our comprehension of intentional action and its many implications, including social and legal ones. This is because, as we have tried to show, the notion of *pattern of action* itself emphasizes the bonds that living beings maintain among themselves in the different moments of action dynamics.

Therefore, we understand that the main problem of the philosophy of action, that of characterizing intentional actions and distinguishing them from reflexive acts, can be better understood when considered in terms of the qualitative aspects of the forms and patterns of action of evolutionarily origin that are shared by organisms.

Acknowledgments

The author thanks CNPq and FAPESP, Grant 2016/50256-0.

References

Ashby, W. R. (1962). Principles of the self-organizing system. In: von Foerster, H., and Zopf Jr., G. W. (Eds.), *Principles of Self-Organization* (pp. 255–278). Oxford: Pergamon.

Atlan, H. (1979). *Entre le cristal et la fumée*. Paris: Editions du Seuil.

Chisholm, R. M. (1966). Freedom and action. In: Keith, L. (Ed.), *Freedom and Determinism*. New York: Random House, pp. 11–44.

Damasio, A. R. (2006). *Descartes' Error*. New York: Random House.

Descartes, R. (1998). *Discourse on Method and Meditations on First Philosophy*. Translated by Donald A. Cress. Indianapolis: Hackett.

Dupuy, J. P. (2000). *On the Origins of Cognitive Science: The Mechanization of the Mind*. Cambridge, MA: MIT Press.

Juarrero, A. (2002). *Dynamics in Action – Intentional Behavior as a Complex System*. Cambridge, MA: MIT Press.

Kelso, J. A. S., and Haken, H. (1995). New laws to be expected in the organism: synergetics of brain and behaviour. In: Murphy, M. P., and O'Neill, L. A. J. (Eds.), *What Is Life? The Next Fifty Years*. Cambridge, MA: Cambridge University Press, pp. 137–160.

Kim, J. (1999). Physicalism. In: Wilson, R. A., and Frank, C. K. (Eds.), *Encyclopedia of Cognitive Science*. Cambridge, MA: The MIT Press, pp. 645–647.

Maini, P. K., and Othmer, H. G. (2000). *Mathematical Models for Biological Pattern Formation*. New York: Springer-Verlag.

Ryle, G. (1949). *The Concept of Mind*. London: Hutchinson.

Schrödinger, E. (1944). *What Is Life? The Physical Aspect of the Living Cell*. Cambridge, MA: Cambridge University Press.

Souza, G. M. (2007). As Formas na Natureza. In: *Caderno de Ensino de Ciências*. São Paulo: Pro-Reitoria de Graduação da Unesp/Páginas e Letras Editora, pp. 49–67.

Visscher, P. K. (2003). How self-organization evolves. *Nature*, 421, p. 20.

von Foerster, H. (1960). On self-organizing systems and their environments. In: Yovits, M. C., and Cameron, S. (orgs.), *Self-Organizing Systems*. Oxford: Pergamon, pp. 1–19.

10

COMPLEXITY AND INFORMATIONAL PRIVACY

A study in the systemic perspective

João Antonio de Moraes and Maria Eunice Quilici Gonzalez

Introduction

Privacy is characterized in traditional studies as personal information accessible only to an individual, or to someone whom the individual deems reliable (Warren and Brandeis, 1890; Schoeman, 1984; DeCew, 2006; among others). From this point of view, the topic of privacy involves notions such as subjectivity, autonomy, and intimacy, among others, and constitutes a problem, for example, when personal information is accessed and/or disclosed without the consent of the individual to whom it refers.

With the introduction of new information and communication technologies (ICTs), such as computers, surveillance cameras, and smartphones, especially, those linked to the Internet, the study of privacy has acquired a greater degree of difficulty. Due to the emergence of the Internet, individuals who in the television era were only receivers of information began to be producers, disseminating information through new technologies. At the present time, personal information can be spread rapidly via information networks.

Given the large amount of information about individuals available in the digital environment, it is difficult to find *criteria of relevance* for analyzing privacy in light of the insertion of ICTs in everyday life. This difficulty characterizes what we call *the problem of the analysis of informational privacy*. We argue that analyses of this problem require not only the investigative resources available in philosophy but also require an interdisciplinary approach that involves the adoption of a systemic perspective. We do not intend to propose a solution to this problem, but to present a methodological alternative of investigation that will support the study of privacy within the scope of information ethics.

Information ethics is constituted as a new branch of interdisciplinary research in philosophy that seeks to reflect on moral issues related to the impact of the insertion of ICTs in everyday life (Capurro, 2006, 2010; Floridi, 2008, 2009). This new ethical proposal brings together elements of traditional ethics, but considers that the valuable contributions of philosophers of the past (such as Aristotle, Kant, Mill, and others) were not directed at moral action in the digital context. It is in this context that we propose a method of investigation, inspired by complex systems theory, in order to approach the problem of informational privacy. Other topics discussed in information ethics are private property, censorship, digital identity, accessibility, responsibility, and ubiquitous computing. As we shall explain, this method allows us to analyze complex phenomena in various spatial-temporal dimensions by identifying patterns that unite them.

In the next section, we introduce central concepts of the complex systems perspective as they apply to the investigation of informational privacy. In the third section, the topic of privacy is made explicit in the context of information ethics. In the fourth section, we develop an analysis of informational privacy in light of the systemic perspective.

Basic concepts of systemics

Among the properties that characterize systemics (i.e., the complex systems perspective), it is relevant to highlight the interdisciplinary *method* of investigation that includes several informational dimensions in the study of events, situations, or objects. In this perspective, cooperation among fields like philosophy, biology, physics, ecology, and sociology, among others, is fundamental to the search for common informational patterns that identify organisms, situations, and objects, without restricting their specificity. We understand that in the study of privacy, this perspective assists in the identification of properties shared by individuals at various levels, and in delimiting what a specific individual or group can consider private.

From the systemic perspective, groups are analyzed as systems while individuals are their elements. According to Bresciani Filho and D'Ottaviano (Chapter 3, this volume):

> A system may be initially defined as a unitary entity of a complex and organized nature, made up of a set of active elements which maintain partial relations between themselves; a system also has characteristics of invariance in time that guarantee its identity. Thus, a *system* is a non-empty set of elements which form a partial structure, with functionality.

Complex systems are informational and materially open; they exchange information, energy, and matter with the environment in which they are inserted. Moreover, systems of this type – for example, living beings and society, among others – are sensitive to the variations of the environment that surround them. Such variations can lead to abrupt and unexpected changes in their elements and even in the whole system. In contrast, informationally closed systems are stable and, while they last, exhibit predictable behavior; an example is a thermostat, which exchanges energy with the environment but does not change its function over time.

In addition to the interdisciplinary nature of systemics research, the following topics are particularly relevant to our proposal for the analysis of informational privacy:

- Self-Organization;
- The principle of emergence;
- The principle of non-linearity;
- Control and order parameters;
- The hologrammatic principle;
- The principle of recursive organization.

Self-Organization is a key feature of complex systems. It is a process of spontaneous organization that occurs among elements of different natures, without the presence of a central coordinator (internal or external) or an absolute controlling center. According to Debrun (Chapter 1, this volume):

> There is Self-Organization every time that the appearance or the restructuring of a form, throughout a certain process, is due to the process itself – and to its intrinsic characteristics – and, only to a lesser degree, to its initial conditions, to the interchange with the environment or to the casual presence of a supervising instance.

Self-organized processes can be classified into primary and secondary, depending on the degree of dependence on their constitutive history. Primary Self-Organization belongs to systems that have abruptly cut back their connections with the past, creating significant changes in their new identity (Debrun, Chapter 1, this volume). Secondary Self-Organization occurs through adjustments of the relations established between the elements of systems, which are constituted during a long-term learning process that allows for the *emergence* of stable qualities that distinguish the identity (Debrun, 2009).

Emergence, supposedly present in the processes of Self-Organization (primary and secondary), is a result of interactions among the elements of the system. This interaction produces *order parameters* that are manifested as informational patterns at the macroscopic level. As stressed by Haken (2000), once established, these parameters enslave the behavior of the elements on the micro level that gave rise to them. They are related to the qualities developed from the interactions among the elements of the system and are not reducible to the mere sum of these elements. In addition, emergent properties are not limited to the physical level but present relational properties embodied in the trajectory of the system's elements.

In the emergency and feedback dynamics of certain systemic properties, *control parameters*, which sustain order parameters, are also present. Figure 10.1, taken from Lewin (1992), illustrates some aspects of this dynamic.

In Figure 10.1, the control parameters are represented in the lower part by the thin arrows; they act on the ways that the elements of the systems relate with each other. The effects of the control parameters on the elements might result in the emergence of the order parameters, as illustrated in the upper part of the figure.

As indicated, the order parameters, once created, feedback on the elements that gave rise to them, influencing interactions among the elements and reinforcing their control parameters (Gonzalez and Haselager, 2007). An example of the presence of control and order parameters in the dynamics of a system can be illustrated through a swimming competition. In this situation, the control parameters would be pool size, water temperature, diving board location, number of swimmers, and other factors that constitute the competitive situation and limit the action of competitors. On the other hand, order parameters can be illustrated by the emergent order manifested on the swimmers' movements, influencing, for example, their speed and style (Moraes, 2014, pp. 117–118).

Although control and order parameters influence the action of the system's elements, the elements themselves do not automatically dissolve and lose their identity in the systems of which they are part; individuals in a society, for example, can promote a revolution in situations when control parameters reach a critical value, altering the prevailing order parameters without, necessarily losing their identity (Haselager and Gonzalez, 2004).

From the systemics perspective, the relation between elements and systems is characterized by the *hologrammatic principle*, which, according to Morin (2005, p. 181), prescribes that, in a sense, the part is in the whole and the whole is in the part. Thus,

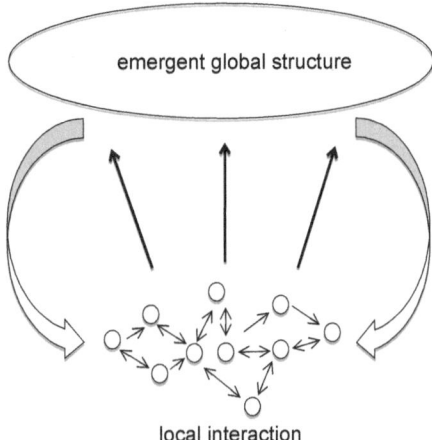

FIGURE 10.1 The interaction between control and order parameters (adapted from Lewin, 1992, p. 13).

for example, human society is present in individuals through laws and the institutions of language and culture; the individual, in turn, is part of the society, acts in its constitution, and contributes to its identity and dynamics. What is at stake here is the process of the emergence of global properties that retroact over the individual parts through the *principle of recursive organization*.

According to Morin (2005, p. 182, our translation), the *principle of recursive organization* prescribes that the effects and products present in an organization are necessary for its own causation. It is in this sense that, "[...] a society is produced by the interactions between individuals and these interactions produce an organizing whole that retroacts on individuals to co-produce them as human individuals". As we will see below an *accelerated recursive organization* is one of the factors responsible for a primary Self-Organization process that seems to be producing a radical break in new conceptions of privacy.

From the characteristics of complex systems indicated earlier, we will try to make explicit our hypothesis according to which the systemics approach provides a method that assists the study of privacy in the informational context, allowing an understanding of the recursive dynamics of society/individual. From this perspective, privacy will be analyzed as a control parameter present in the relationships among individuals in their specific contexts, governed by the order parameters of present-day information ethics.

Privacy from the perspective of information ethics

As indicated, information ethics is a branch of information philosophy that deals with the directions ICTs are taking in the contemporary industrial society. In terms of the analysis of privacy, illustrations of how new ICTs can be a problem are the following: (1) social networks on the Internet, which make possible the acquisition and dissemination of information about individuals, without their being aware of it; (2) surveillance cameras, which cannot only restrict individual actions but can also record information about individual and collective habits, without the consent of individuals.

The new possibilities of interaction provided by ICTs can promote a sense of dependence on being *online*. Moreover, even if the individual does not want to be online most of the time, this feeling remains due to the dissemination in the daily life of information devices such as cameras, credit cards, etc. In this situation, the questions arise: What are the ethical implications of the insertion of informational technologies into everyday action? What are the implications of such an insertion for current notions of privacy?

The analysis of privacy poses itself as a problem in the scope of ICTs due to the great existing capacity for capturing and disseminating information about individuals, and this problem requires a criterion of relevance as the starting point of the investigation. To further complicate this scenario, the notion of privacy adopted by common sense has been altered by the insertion of informational technologies in daily life.

We understand that a process of primary Self-Organization is underway in the field of ethics, generating radical ruptures, for example, in the traditional (Western) conception of privacy. The intimate concept of a "private life", belonging only to the subject and accessible only to the individual or to those whom he or she deems reliable, is changing for users of the new ICTs. This is due to interpersonal interactions carried out through interfaces that occur mainly by means of online social networks such as Facebook, Instagram, and Twitter. Users of these platforms fill in their profiles with information about birthdays, places they frequent, marital status, sexual preference, friendships, etc. In this context, the traditional notion of privacy seems to be inadequate for understanding the limits of what is admitted as private because people expose themselves on the Internet without experiencing the feeling of invasion of privacy.

One of the factors responsible for this abrupt rupture between public and private life seems to be the *familiarity* that users have acquired with the routine use of ICTs, which facilitates the exposure of personal information in online social networks; the frequent use of such technologies induces people to continue using them without much questioning. The filling in of personal information on online social networks is, much of the time, done mechanically. Factors such as age and ignorance of how such technologies work also contribute to the provision of personal information to social networks in an "automatic" way.

The interaction between individuals made possible by online social networks makes it difficult, paradoxically, to identify the *limits* of what is considered to be subject to individual protection. Thus, an initial difficulty in the analysis of informational privacy is to delimit a context in which information conceived in the past as private is now made explicit and accessible in virtual environments. This difficulty points to a kind of rupture between the past and future of a system (a rupture that is typical of primary Self-Organization); in this case, a rupture occurs between the notion of privacy of a generation unfamiliar with the new ICTs and the ideas about privacy developed by a generation raised with these technologies.

The difficulty in identifying the boundaries between the public and private spheres can also be a responsible factor in the constitution of a "surveillance society". This expression is used to characterize the sensation of observation generated by the presence of ICTs in many areas of life. In contrast to the "rebellious generation" of the 1960s (who revolted against control and surveillance), the new generations seem to be divided into two groups with respect to the presence of cameras in public environments: the indifferent ones and those who feel protected.

The "surveillance society" is currently marked by the presence of *ubiquitous computing*, a term introduced by Weiser (1991) to characterize information processing systems that are disseminated in the environment, capturing, storing, and transmitting information all the time. A central feature of ubiquitous computing is that it is spread out, without a specific controlling center, and most often acts without individuals' attention. Examples of this kind of computing are security cameras, barcode badges, biometrics systems, and the frequency registers found in environments such as stores and offices.

We understand that ubiquitous computing is one factor responsible for the maintenance of the "surveillance society", because much of the information acquired through this type of technology refers to the particular *habits* of individuals; without their being aware of it, information is being collected that may be used to jeopardize their privacy. With the use of ICTs, supermarket managers, for example, know what individuals consume, as do bank managers, who have data from financial transactions, or companies that record conversations and phone calls, sometimes without authorization.

In summary, the impacts of ICTs on everyday life indicate a rupture in the conception of privacy, insofar as they result in a significant alteration of what is meant by privacy in the "information age". To understand the limits of what can be considered private has become a challenge after the widespread insertion of ICTs into our daily activities. A large amount of information available on the individual in online social networks, and via ubiquitous computing, indicates the difficulty of establishing a criterion of relevance for analyzing privacy in the informational context. In the next section, we propose a systemic approach for investigating this difficulty.

Informational privacy in the light of systemics

From the systemic perspective, privacy can initially be analyzed as an emergent property of shared relations among individuals and groups, which has a greater or lesser degree of expansion due to characteristics related to the location of each individual (Moraes, 2014, p. 113). From this perspective, privacy can be studied as a control parameter conventionally constituted from information that is meaningful for the individuals inserted in certain groups in specific contexts.

The geographic location of the individual is relevant to the analysis of privacy because his/her conduct (in both the "real" world and the virtual network) reflects aspects of his/her environment. However, it is an individual's informational location – his or her belonging to a group – and not necessarily a geographical location, that provides a basis for the establishment of a criterion of relevance for privacy analysis. It is, initially, the situated and incorporated individual that makes possible the establishment of what one considers private; without an individual being situated and incorporated, there is no privacy. But the conditions that we deem sufficient for delineating what can be considered private will ultimately depend on the values conventionally established in a group, which function as control parameters for the informational system that defines the group.

The hologrammatic property mentioned earlier is expressed here as an indication that explicit privacy in the group (the whole) is reflected in the individuals (the parts). In this regard, Mainzer (1997, p. 313) makes the following observation about the role of the individual in the dynamics of society:

> From a macroscopic viewpoint we may, of course, observe single individuals contributing with their activities to the collective macrostate of society representing cultural, political, and economic order ('order parameters'). Yet, macrostates of a society, of course, do not simply average over its parts. Its order parameters strongly influence the individuals of the society by orientating ('enslaving') their activities and by activating or deactivating their attitudes and capabilities. This kind of feedback is typical for complex dynamical systems.

It should be emphasized that the notion of privacy outlined here presupposes the existence of a process that involves a certain degree of Self-Organization, since it is not imposed by an absolute controlling center, but constitutes the very dynamic of interactions between individuals. Although *Google*, among others, can control the information available on the network, from the systemic perspective there is still a space for spontaneous interaction among users that allows them some degree of freedom.

The analysis of what deserves to be protected in the systemic context can be carried out in terms of two levels: (1) the way in which privacy is dynamically conceived by the group, or (2) the individual characteristics that can act as control parameters in the emergence of order parameters that constitute different conceptions of privacy. As we pointed out individuals can bring about significant changes in traditional views of morality, politics, and, in the same vein, privacy, when existing control parameters wear out or reach a limit. In other words, an individual can act, both from a geographical and an informational location, as a precursor of a new conception of what deserves to be protected. If others share the individual's conception, it may gain strength and possibly eventually change the current norm. If the *principle of accelerated recursive organization* applies to this situation, new control parameters will quickly be established and consolidated in the construction of a new conception of privacy.

One advantage of adopting the systemics perspective in the study of privacy is that it allows one to understand the process of primary Self-Organization underlying the ruptures in the dynamics of society. In the case of the notion of privacy, this method allows one to understand the different temporalities to which individuals of the same time period are submitted. The principle of accelerated recursive organization functions in a truly accelerated way in young users of digital technologies who quickly learn to manipulate artifacts by changing their habits, which will function as the cause of their own order, changing not only their conceptions of the world but their own notions of privacy and identity. On the other hand, in the older generations, this dynamic occurs (if at all) at a slower pace, allowing long-term relationships to remain unchanged with respect to conceptions of privacy.

In summary, in light of the systemic perspective, criteria of relevance adopted for the analysis of informational privacy may be stated in terms of:

a   *Context*, geographical and/or informational, in which the exposure of personal information by individuals occurs;
b   *Self-organizing dynamics*, which acts in the formation of control and order parameters that maintain the worldview of individuals; and
c   The *principle of accelerated recursive organization*, which acts differently in different generations coexisting in the same era.

We understand that in each context, conditions (a), (b), and (c) provide tools for distinguishing what is considered private from the properties that individuals consider worthy of being protected. Thus, for example, while for a particular group of individuals (mainly those who have greater proximity to the use of ICTs) the access to information by others (even unknown others) does not constitute an invasion of their privacy; for members of another group, the simple use of biometrics in a library, by recording certain habits and preferences, can cause a sense of invasion of privacy.

Final considerations

In this chapter, we have discussed the problem of informational privacy in the light of the insertion of ICTs in our daily life. As we have stressed, these technologies enable individuals to become information producers and producers of information about other individuals. In this situation, privacy becomes a difficult problem of analysis, because, in the digital domain, once the information is disseminated one does not have total control over it; ICTs' potential for information sharing does not involve a single control center. The problem of analyzing privacy in this context is, therefore, mainly one of finding criteria of relevance for identifying what individuals consider private.

Given the various factors concerning the problem of informational privacy, we have outlined a systemic approach according to which privacy is conventionally constituted by individuals or groups from properties that are considered worthy of being protected. In this analysis, disparate notions of privacy related to differences of context (for example, generational differences) are understood as mainly due to the *principle of accelerated recursive organization*.

We claim that the analysis of privacy according to the systemics perspective has the advantage of also accounting for virtual reality, here understood as the networks of relations which make up the informational location of individuals.

One question that can be raised about the suggested systemic perspective on privacy is this: How is it possible to avoid conceptual relativism in the proposed systemic analysis of privacy? We believe that perspectivism should be adopted to prevent conceptual relativism. Perspectivism, as proposed by Peterson (1996), is a methodological stance through which multidimensional explanations can be constituted, depending on the task of investigation. According to this viewpoint, the development of possible levels of analysis occurs in a search for a common pattern that identifies a system that has several dimensions. In this sense, a systemic approach within the perspectivist perspective can be understood as a methodological stance according to which the identification of a common pattern is to be found in the relations between part and whole on the various levels of analysis (individual/individual, individual/group, group/group).

Another difficulty with the problem of analysis of informational privacy is the constitution of *hybrid beings*, which has emerged as a consequence of individuals interacting with ICTs (Floridi, 2005, 2014; Moraes and Andrade, 2015); they grow up within the informational context, as is the case with many individuals born after 1996 in industrialized societies. For these individuals, contact with ICTs does not cause a feeling of strangeness.

The principle of accelerated recursive organization allows us to understand the familiarity with technologies present in the interaction of hybrid beings, and how this familiarity promotes the tacit approval of such technologies. This approval is due to the immediate benefits presented on first impression, the damages only being perceived over time (Quilici-Gonzalez et al., 2010; for example, the disclosure of information about an individual without his or her consent).

We also believe that the familiarity and tacit acceptance of ICTs by individuals in their daily lives contributes to their adoption of practices of transparency of information with regard to their habits. Capurro (2005, p. 42) considers that the traditional concept of privacy is being replaced by the notion of "*Be transparent!* and then you are a good citizen". In this context, new questions arise: How should one balance the transparency/privacy relationship, given that the practice of transparency is ostensibly implemented for the sake of public safety? Is the tendency towards a "transparent" society inevitable? Is the end of privacy only a matter of time?

Unlike Kronenberger (1964), we feel that privacy has not yet taken its final blow. Even with the difficulties engendered by the principle of accelerated recursive organization as seen in the insertion of ICTs into contexts of human action, we understand that if one begins to reflect on the impacts of such insertions, it will still be possible to preserve multifaceted privacy, even in the virtual environment. In this context, the role of the philosopher is of great importance, since a critical view of the moral and political consequences of the action of individuals is necessary.

Finally, privacy stands out in current discussions as one of several topics that have become difficult to analyze when placed in the informational context. There are more issues than answers, due to the novelty that the uses of ICTs have brought to this subject. We understand that discussions about privacy in the virtual environment and the interaction between individuals and ICTs should be key issues in the interdisciplinary philosophical agenda of new directions that philosophy has taken in the "information age".

Acknowledgments

The authors thank FAPESP (Process 6/50256-0) and CNPq (Process 310931/ 2015-0) for supporting this research, and colleagues from the UNICAMP and UNESP for suggestions and criticisms.

References

Capurro, R. (2005). Privacy. An intercultural perspective. *Ethics and Information Technology*, 7, pp. 37–43.

———. (2006). Towards an ontological foundation of information ethics. *Ethics and Information Technology*, 8(4), pp. 175–186.

———. (2010). Desafíos téoricos y practicos de la ética intercultural de la información. *E-Book do I Simpósio Brasileiro de Ética da Informação* (pp. 11–51). João Pessoa: Idea.

Debrun, M. M. (2009). *Brazilian National Identity and Self-Organization*. Campinas, Brazil: UNICAMP University Press. CLE Collection

DeCew, J. (2006). Privacy. *Stanford encyclopedia of philosophy*. Available at: http://plato. stanford.edu/entries/privacy/. Accessed on August 27, 2017.

Floridi, L. (2005). The ontological interpretation of informational privacy. *Ethics and Information Technology*, 7, pp. 185–200.

———. (2008). *Information ethics, its nature and scope*. Available at: www.philosophyo finformation.net/publications/pdf/ieinas.pdf. Accessed on August 01, 2017.

———. (2009). *The information society and its philosophy: introduction to the special issue on "the philosophy of information, its nature and future developments"*. Available at: www. philosophyofinformation.net/publications/pdf/tisip.pdf. Accessed on August 01, 2017.

———. (2014). *The Fourth Revolution: How the Infosphere Is Reshaping Human Reality*. Oxford: Oxford University Press.

Gonzalez, M. E. Q., and Haselager, W. F. G. (2007). Mechanicism and autonomy: what can robotics teach us about human cognition and action? In: Gonzalez, M. E. Q., Haselager, W. F. G., and Dror, I. E. (Eds.), *Mechanicism and Autonomy: What Can Robotics Teach Us about Human Cognition and Action? Pragmatics & Cognition*, 15(3), special issue, pp. 407–412.

Haken, H. (2000). *Information and Self-Organization*. New York: Springer-Verlag.

Haselager, W. F. G., and Gonzalez, M. E. Q. (2004). Consciousness and agency: the importance of self-organized action. *Networks*, 3(4), pp. 103–113.

Kronenberger, L. (1964). *The Cart and the Horse*. New York: Alfred A. Knopf.

Lewin, R. (1992). *Complexity: Life at the Edge of the Chaos*. New York: Macmillan Publishing Company.

Mainzer, K. (1997). *Thinking in Complexity: The Complex Dynamics of Matter, Mind, and Mankind*. New York: Springer.

Moraes, J. A. (2014). *Implicações éticas da "virada informacional na Filosofia"*. Uberlândia: EDUFU.

Moraes, J. A., and Andrade, E. B. (2015). Who are the citizens of the digital citizenship? *International Review of Information Ethics*, 23(11), pp. 4–19.

Morin, E. (2005). *Ciência com consciência*. Rio de Janeiro: Bertrand Russel.

Peterson, D. (Ed.). (1996). *Forms of Representation: An Interdisciplinary Theme for Cognitive Science*. Wiltshire: Cromwell Press.

Quilici-Gonzalez, J. A., Kobayashi, G., Broens, M. C., and Gonzalez, M. E. Q. (2010). Ubiquitous computing: Any ethical implications? *International Journal of Technoethics*, 1, pp. 11–23.

Schoeman, F. (Ed.). (1984). *Philosophical Dimensions of Privacy: An Anthology.* Cambridge: Cambridge University Press.

Warren, S., and Brandeis, L. (1890). The right to privacy. *Harvard Law Review,* 4, pp. 193–220.

Weiser, M. (1991). The computer for the 21st century. *Scientific American,* 265(3), pp. 94–104.

11

AN UNDERSTANDING OF SOLIDARITY ECONOMY IN THE LIGHT OF SELF-ORGANIZATION THEORY

Renata Cristina Geromel Meneghetti

Introduction

This chapter focuses on the constitution and operation of solidarity economy enterprises (SEE) and analyzes them in light of Michel Debrun's theory of Self-Organization. Solidarity economy is understood as a "group of economic activities – production, distribution, consumption, saving, and credit – organized and executed cooperatively by workers in a collective and self-managed manner" (Brasil, 2006a, p. 11, my translation). A variety of enterprises can be included in this category, such as cooperatives, associations, exchange clubs, worker-recovered enterprises (a form of workers' self-management), solidarity finance organizations, and informal groups. These enterprises are characterized by some form of economic activity, and by cooperation, solidarity, and self-management.

Self-Organization, as opposed to hetero-organization (which suggests fully planned systems), refers to a system that can "be the genesis of its own being" (Debrun, Chapter 2, this volume). According to this author, a self-organized process is constituted or restructured from the interaction between distinct or semi-distinct elements; and it can happen on two levels: the first is primary Self-Organization (referring to the creation of an organization), and the second is secondary Self-Organization (focusing on the restructuring of an organization) (Debrun, Chapters 1 and 2, this volume).

According to Andrade (2011), an organization can manifest itself in different contexts of reality:

> Generally, an 'organization' has been represented, in an abstract or formal manner, as a structure, a group of elements, and the relationships between these elements. It is up to the various areas of knowledge to fill the

components of this structure with some empirical content, such as elements of a specific type, relationships, laws, and principles operating in real systems.

(Andrade, 2011, p. 79, my translation)

The word *context* is understood here as Andrade (2011) interprets it: "the field, the connection between a group of possible circumstances (the characteristics of the domain), and a group of modes of behavior adjustable to these circumstances (the characteristics of the image); the domain and image of habitual relationships" (Andrade, 2011, p. 92, my translation).

Therefore, in this chapter, I propose understanding SEEs as Self-Organizations. First, I argue that they can be understood as primary Self-Organizations (in the moment of the establishment of these SEEs). Second, I suggest that SEEs have the potential to constitute secondary Self-Organizations, since learning is a fundamental aspect of their evolution in the context of solidarity economy; learning is also what allows for adjustments and the inclusion of innovations in self-organized systems (Broens, 2004).

On solidarity economy

Capitalism is a mode of production based on the right of the individual to own property and capital, and on individual freedom for competition. As a result, there is a progressive increase in inequality, dividing society into winners and losers (Singer and Souza, 2000).

According to Singer (2002), capitalism has apparently become dominant, common, and natural; a market economy that is not competitive seems unthinkable. There are those such as Singer, however, who criticize the competitive economy for its social effects. In the author's words, "The defense of competition puts only the winners in evidence; the fate of the losers is left in the shadows. [...] In a capitalist economy, winners accumulate advantages, and losers accumulate disadvantages for future competitions" (Singer, 2002, p. 8, my translation).

Kliksberg (2002) affirms that along with the neoliberal hegemony of the last two decades of the twentieth century, it has been possible to notice a deterioration in the social conditions of great parts of the population, especially in underdeveloped countries. Even though the supporters of this policy argue that free markets can produce greater efficiency and welfare for all, reality displays an immense concentration of income and an increase in poverty and social exclusion.

According to this author, problems such as low educational level, lack of access to health services, high levels of unemployment and job insecurity, an increase in crime rate, the destruction of the family, among other factors, produce a perverse circle of exclusion that only tends to generate more poverty, making the whole social situation more and more untenable.

Singer and Souza (2000) point out that achieving an egalitarian society requires a solidarity economy in which participants cooperate rather than compete. Solidarity economy bases itself on the collective ownership of capital and on individual freedom; it gives rise to a single class of workers and results in solidarity and equality.

Such an economy springs from the context of popular economy, which can be understood as

> the set of economic activities and social practices developed by popular sectors to guarantee, by using their own available workforce and resources, the satisfaction of basic needs, both material and immaterial.
>
> *(Icaza and Tiriba, 2003, p. 101, my translation)*

According to Kruppa (2005), solidarity economy proposes equality of conditions and the right to be different. The principle of equality of conditions seeks to eradicate a hierarchical society by supporting democratic relationships in which differences do not generate inequalities. It is an economy that not only includes differences but also allows for the practice of these differences. Therefore, solidarity economy is understood as "a set of economic activities – of production, distribution, consumption, saving and credit – organized and performed with solidarity by workers under a collective and self-managed form" (Brasil, 2006a, my translation).

Solidarity economy has four important characteristics: cooperation, self-management, economic viability, and solidarity. Cooperation is understood as the existence of common interests and objectives, and as the solidary responsibility for overcoming adversities. Self-management refers to participative practices for managing a group's activities. Economic viability means the joining of efforts to enable the group's collective initiatives. Solidarity is a concern for what is just, in order to provide for the welfare of the workers and consumers involved in the process.

In Brazil, for the last fifteen years, solidarity economy has been growing as a social movement. According to the Brazilian Solidarity Economy Forum (FBES), this type of economy has been a

> result of the organization of workers in the construction of economic and social practices based on relationships of solidary collaboration, inspired by cultural values placing the human being as the subject and the purpose of economic activity, instead of the private accumulation of wealth in general and capital in particular.
>
> *(FBES, 2006, p. 3, my translation)*

However, for a solidarity enterprise to work well, unity and interest are required, whereas conflicts, competition, and disputes between workers must not exist. This calls for a collective reeducation of workers, for them to help each other, and for decisions to be made collectively. Workers must understand that they are all equal owners and that each one has an equal deciding power over the matters of the enterprise; in other words, that every worker's vote for the decisions made inside the enterprise carries the same weight. Therefore, solidarity economy can only happen if it is equally organized by those associating to produce, sell, purchase, and save. In it, everyone has the same share of the capital and, consequently, the same voting right in every decision. In a solidarity enterprise, members also must decide collectively, in an assembly, if the withdrawals (the salaries) will be equal or differentiated.

Another distinction refers to the way in which a solidarity enterprise is administered. For Singer and Souza (2000), type of management seems to be the main difference between capitalist and solidarity economies. In a capitalist economy, there is *hetero-management*, in which a hierarchical administration occurs with successive levels of authority and information flows from the bottom to the top while orders flow from the top to the bottom. Furthermore, according to Singer and Souza (2000, p. 21), from our school years, we are taught to "obey and fear the 'superiors', [...] in an educative process that continues through our whole lives" (my translation). These authors believe that we are not naturally inclined towards hetero-management – rather, we are made used to it. On the other hand, in solidarity economy there is self-management, in which a democratic administration occurs without a managerial hierarchy, and where information flows from the top to the bottom and orders flow from the bottom to the top (decisions are taken in assemblies, which can occur whenever there is a need for one).

Lechat and Barcelos (2009) point out that *autogestão*, the Portuguese term for *self-management*, means literally to administer, to manage oneself – from the Greek *autos* (self) and from the Latin *gest-o* (manage) – and it is used to define groups that organize themselves without people in official leadership positions. The principle of self-management derives, therefore, from the philosophical and political notion that all people are able to organize themselves without rulers.

According to the same authors, self-management, when referring to a form of organization of collective action, possesses a multidimensional aspect (social, economic, political, and technical). Self-management has a social dimension because "it must be perceived as the result of a process able to articulate actions and results that are acceptable by all individuals and groups depending on it" (Lechat and Barcelos, 2009, my translation). Its economic aspect comes from the social relations of production, valuing work over capital. Its political aspect is based on systems of representation with values, principles, and practices that create the conditions

> to make the decision-making process the result of a collective construction using shared power (of expressing opinions and of deciding), in order to guarantee a balance of forces and a respect for the different actors and social roles of everyone inside the organization.
>
> *(Lechat and Barcelos, 2009, p. 2, my translation)*

The technical aspect is what allows us to think of another form of division and organization of work.

The biggest threat to self-management is the disinterest of the worker-members, because self-managerial practices risk being disrupted by the principle of least effort. Thus, for self-management to succeed, it is required that all members to keep informed about what is happening in the enterprise and about the available options for solving each problem, because a solidarity enterprise belongs to every member.

Self-management aims to make solidarity enterprises economically productive, and to allow for the creation of democratic and egalitarian hubs for interaction. Not only economic efficiency, but also human development, can be achieved through this form of management: "participating in the discussions and in the decision-making of the collective to which one belongs both educates and raises awareness, making the person feel more accomplished, more self-confident and safer" (Singer, 2002, p. 21, my translation).

Therefore, solidarity economy is a form of wealth production, consumption, and distribution – an economy – based on valuing the human being, not the capital, in an associative and cooperative manner, directed towards production, consumption, and commercialization of goods and services in a self-managed way. Solidarity economy aims at human freedom through work, in a process of economic democratization that creates an alternative to the alienating and wage-based labor relations of capitalism: "its basic final purpose is not to maximize the profits, but the quantity and quality of work" (Singer, 2002, p. 4, my translation).

As noted earlier, SEEs may take a variety of forms, both formal and informal, and are characterized by cooperation, solidarity, and self-management. In this chapter, I will discuss a small SEE, a collective carpentry shop that is run by women (which, due to its small size, cannot, strictly speaking, be considered a cooperative).

Veiga and Fonseca (2001) describe a cooperative as follows:

> A nonprofit voluntary association, with economic purposes, of at least twenty people executing the same activity towards common objectives, in which members contribute equally to the formation of the necessary capital through the acquisition of shares, and accept sharing equally the risks and benefits of the enterprise. It is regulated by the democratic principle of one person, one vote. Surpluses are distributed according to the amount of work of each member.
>
> *(Veiga and Fonseca, 2001, p. 39, my translation)*

A regulating law in Brazil – the *Lei Cooperativista 5.764* of December 16, 1971 – characterizes a cooperative as "a society of people with a specific legal nature and form, of a civil nature, non-susceptible to bankruptcy, constituted to perform services for its members" (Veiga and Fonseca, 2001, p. 39, my translation).

Matsuda (2008) describes the formation of cooperatives as the technical and political pursuit of the balance between the social and the economic. For these authors, cooperatives are a set of collective organizations with the fundamental goal of generating work and income, the conditions of work and life.

According to Singer (2002), modern cooperatives arose as an alternative to capitalism at the peak of the industrial revolution, when the labor movement was going through a period of strong tension. Nowadays, the return to this form of organization occurs due to rising unemployment rates and the increasing precariousness of labor relations. This current situation is a result of the neoliberal policies, technological developments, and economic and financial globalization, all of which have taken hold of the world in general, and Latin America in particular, over the last two decades (Singer, 2002).

On learning in the context of solidarity economy

In the context of solidarity economy, learning is paramount, since it is through education and training in solidarity economy that people can guarantee, as a result of pedagogical practices, both their survival and the improvement of their life conditions, thus, encouraging the development of social protection networks (Brasil, 2006b). Moreover, through education and training, the organization of workers around social-economic projects valuing work (not capital) can be strengthened. Therefore, educational processes inspired by solidarity economy point in the direction of a new sociability, a new society, a new form of life production.

Another challenge for education is creating a collective inquiring spirit – able to include every actor participating in the training process – both for the unveiling of the world, as well as for the searching of ways to favor political, economic, social, and cultural transformations (Brasil, 2006b, p. 17, my translation).

Brandão (1986) sees education, in relation to its social context, as the condition for the permanent recreation of culture itself. From the perspective of the individual, education is the condition for the creation of the individual, and it occurs through the exchange of knowledge between people. According to the same author, *learning means constituting oneself from the organism into an individual, accomplishing the passage from nature to culture.* First, there was a common knowledge which became wise and scholarly; scholarly knowledge then defined the common sense knowledge (from whence it arose) as *popular.* Scholarly knowledge defined itself in its own form, centralized and associated with education specialists, while popular knowledge was increasingly regarded as diffuse knowledge pertaining to the way of life of a subordinate strata of society. In Brandão's view,

> The *knowledge of the community* becomes the *knowledge of the fractions* (classes, groups, peoples, tribes), subalterns in an unequal society. In a primary sense, forms – whether immersed in other practices or not – through which the popular classes' knowledge is transferred between groups and people, are their *popular education.*
>
> (Brandão, 1986, p. 26, my translation)

This divide between so-called scholarly and popular knowledge brings about the exclusion of the oppressed, of the subordinate strata of an unequal society. It is in opposition to this process that popular education rises. Popular education is an engaged and participative education guided by the perspective of achieving all of a people's rights (Brandão, 1986). Its main characteristic is using the community's knowledge as the raw material for teaching. It is learning from individuals' own knowledge and teaching from the words and themes that produce their own everyday lives. The process of teaching and learning is regarded as an act of social knowledge and transformation, having a certain political aspect.

Popular education seeks to form individuals who have knowledge of their rights and a consciousness of their citizenship, and to organize the political work towards the affirmation of the individual. It is a strategy of constructing popular participation in order to redirect social life.

The result of this form of education can be observed when individuals can firmly place themselves in their context of interest. Popular education can be applied to any context, but it is more commonly implemented in rural settlements, indigenous villages, and in adult education. It is most prevalent in relation to social movements, for they are the means through which minorities' voices are heard.

On the constitution and functioning of an SEE: the women's collective carpentry shop

Since 1998, a Regional Incubator of Popular Cooperatives (IRCP, *Incubadora Regional de Cooperativas Populares*), established by a public university, has dedicated itself to teaching, research, and extension activities relevant to solidarity economy. Some of these activities are involved in the process of incubating SEEs in different economic activities and locations. This incubating activity began when a group of members of the university established contact with the population of one of the city of São Carlos's poor neighborhoods. Discussions arose from this contact about addressing the needs of the area, particularly with regard to generating income for a population excluded not only from the labor market but from several other conditions inherent to citizenship. In these discussions, the population was presented with the possibility of organizing work collectively, in accord with solidarity economy proposals. Moreover, productive activities that could justify enterprises of this nature were identified, such as activities related to the construction and food industries, sewing, and the cleaning of buildings. From the identification and examination of some of these activities, collective work initiatives were started. This process originated an integrated system of SEEs supported by the São Paulo Research Foundation (FAPESP), in a public policies project which I joined in 2009 as a coordinator of a group responsible for mathematics education. One of these SEEs, a collective carpentry shop run by women from a rural settlement in the southwest region of the state of São Paulo, is discussed later. It has been supported both by the IRCP and by the Housing and Sustainability Group (GHS, *Grupo de Habitação e Sustentabilidade*), which is part of another public university located in a rural area of São Paulo State. The latter group works in the field of sustainability in several dimensions – environmental, social, economic, and political – and also works to develop awareness on the part of citizens that they are agents able to alter their own realities. In what follows, I intend to contextualize this SEE, discussing its formation and its operation.

The women's collective carpentry shop (MCF, *Marcenaria Coletiva Feminina*) is located in a rural settlement in the southwest region of the state of São Paulo. It was created as a part of a larger project, named Inovarural, that was coordinated by the GHS and the IRCP and initiated in January 2003. The activities of the carpentry shop began in 2004 with the construction of items such as windows and doors for the settlement's own houses. After the construction of the houses, an opportunity arose for the cooperative's members to learn a new trade for generating income.

Difficulties presented themselves during the activities of the MCF, some of which related to learning mathematical topics. IRCP researchers noticed the need for a project focusing on mathematics education, in order to aid the members' emancipation in the direction of self-management and to serve as a support instrument for a greater level of control over the production chain and the goods being produced. Basic mathematical concepts, namely arithmetic, algebra, and geometry, as well as financial mathematics and the use of electronic spreadsheets, were among the greatest difficulties.

The data used in the elaboration of the flowcharts representing the production chain of the MCF under discussion are based on active observation, participation in meetings with the incubator responsible for implementing this SEE, and interviews with members of the MCF. In the flowcharts presented later, rectangular boxes represent the processes of the production chain of this SEE, while round boxes indicate the beginnings and ends of the processes.

Firstly, an overall flowchart was first drawn to represent every step in the MCF production chain (Figure 11.1). The flowchart is composed of seven steps, from which sprung seven other sub-flowcharts, one for each of the steps of the production chain. Next, I will briefly present what each of these steps entails.

The MCF's production chain starts with the costumer's order and the designing of the project. Next, the net amount of wood required for the making of the product is calculated. The ending of this process (step 1: sub-flowchart 1) is the definition in cubic meters of the net wood, which is followed by the process of verifying the existence in a stock of gross wood (step 2: sub-flowchart 2).

This verification leads to the amount of gross wood required, and only then can the budget be calculated (step 3: sub-flowchart 3). This calculation results in a budget which may or may not be approved by the customer who hired the

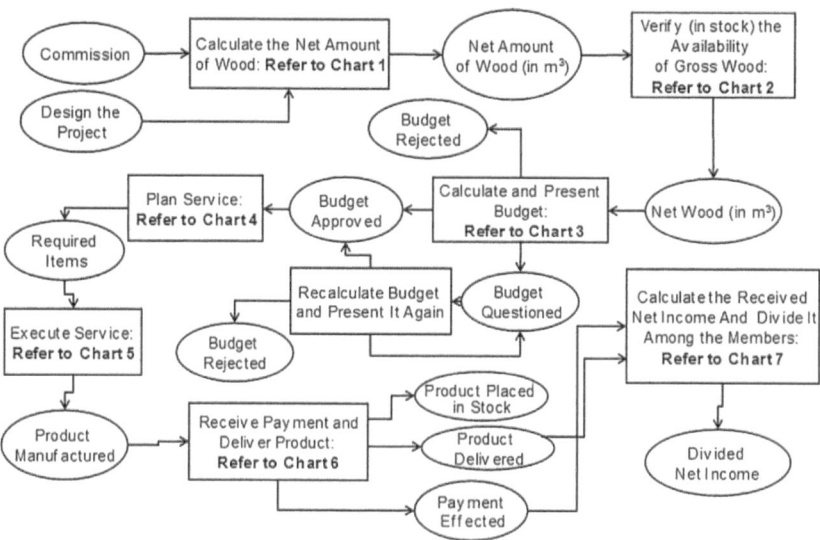

FIGURE 11.1   Flowchart of MCF's overall production chain.

service. If the budget is not approved, the carpenters recalculate the budget and try to negotiate. If rejection on the part of the customer persists, it is the end of the production chain. If the budget is approved, the service is planned (step 4: sub–flowchart 4).

After the planning step, which includes the definition of the necessary items for the fabrication of the product, the next step is the execution of the service (step 5: sub–flowchart 5). This execution results in the manufactured product, which will be delivered in the penultimate step, which involves delivering the product and receiving the payment (step 6: sub–flowchart 6). The end of this process is the delivery of the product and payment by the client. If it is not possible to deliver the product to the buyer, it must be placed in stock. Lastly, once payment is received, the members of the carpentry shop can calculate and divide the net income obtained by subtracting the expenditures from the gross income (step 7: sub–flowchart 7). The end of the production chain occurs when the net income is divided among the members of the SEE that participated in the manufacturing of the product.

Attempting to understand SEEs in light of the theory of Self-Organization

Returning now to the objectives of this research, I suggest that it is possible to correlate solidarity economy with the theory of Self-Organization, considering that SEEs aim at achieving self-management, which implies becoming a self-organized organism. This is because self-management must be present in every process of an SEE's production chain, in every action, and in every decision made. As mentioned before, according to Lechat and Barcelos (2009, p. 2), self-management refers to a form of organization of collective action that has a multidimensional aspect (social, economic, political, and technical).

However, it is important to point out that the Self-Organization of an SEE might not be successful. According to Debrun (Chapter 1, this volume), the simple joining of elements surrounding a cooperativist project does not result in Self-Organization – an interaction between the elements is required, and "not only the strict conditioning of elements with a certain degree of freedom that are possible participants of a process. This condition opposes 'self' to 'hetero' organizations" (Andrade, 2011, p. 83, my translation). Similarly, unity and interest between its members are necessary for an enterprise to become an SEE. The risk of an enterprise not becoming an SEE comes mostly from the non-occurrence of self-management.

In relation to SEEs like the ones focused on in this research, I have observed a beginning in which there is a strong separation between what came before and what comes after. These SEEs are turning points in people's lives. At first, different subjects (similar, however, in their precarious life conditions) get together with a common purpose (that of avoiding hierarchical administrations), attracted by the guiding principles of an SEE: self-management, democracy, participation, equality, cooperation, self-support, human development, and social responsibility (Gaiger, 2004). This form of constitution is similar to the way that Debrun describes self-organized organisms (or forms).

In the SEEs that I observed, I noticed a significant *attractor* in the process of constituting these enterprises: the possibility of changing these people's life conditions. Extracts from members' interviews with members of Women's Collective Carpentry Shop point in this direction, as do comments of members of the Cleaning Cooperative (Cooperativa de Limpeza), another SEE established with support from the IRCP: "we used to suffer a lot of discrimination, my life changed, today I have a car, a completely furbished house" [Member E, Cleaning Cooperative];

> it really helped those of us from the neighborhood, it was a neighborhood really... it used to be a favela, didn't it? [...]. Today we have an income, before we had nothing. People's self-esteem increased, especially women's, who didn't have an income before; there are some people who had never had a job, and with the cooperative's arrival they have their incomes today;
> *[Member F, Cleaning Cooperative]*

> A little extra income to help the family, isn't it? It is a job for us, from the countryside, you know, who couldn't handle the countryside anymore, the hot sun. We can't stand it anymore at our age, can we? And this job here we can stand. [...] It changed a lot, yes, because we used to live in a rickety house, and also our jobs. Time went by really fast for me this year. These two and a half years, time flew by, I didn't even notice. I used to get sick

from time to time, and after I began working here I don't even get sick anymore! It seems stress was making my blood pressure go up, you know, and after, not anymore, we got busy. Our minds got busy … I don't know. It feels good. It is also an extra income, isn't it?

[Carpenter S; Women's Collective Carpentry Shop]

From these statements, it may be seen how the generation of jobs and income, and the possibility of changing one's life, are motivating the members to constitute and maintain the SEE; these incentives thus work as attractors, in accord with the theory of Self-Organization.

In addition, as Gonzalez et al. (2000) suggest, individual changes are reflected in the organization itself in a dynamic and continuous manner, due to what the authors call *"circular causality."* The whole modifies the parts, and the parts modify the whole:

> [...] individuals create several types of social organizations that begin to affect their own personal identities. The individual changes, then, are reflected in the organizations that the individuals create; these, in turn, will affect each one of them, and so forth, in a dynamic and circular manner.
>
> *(Gonzalez et al., 2000, p. 70, my translation)*

Personal identity itself is, therefore, understood by these authors as a Self-Organization process that gradually acquires stability and autonomy.

Broens (2004) affirms that for the theory of Self-Organization, the subject is always a contextualized one. The context, therefore, influences the subject while the subject is influencing the context. There is always an interaction, there is always a context in which the subject belongs. The author also notices that for the theory of Self-Organization, the concept of a subject is defined in the framework of secondary Self-Organization "inasmuch as a self-organized system is able to incorporate innovations in the inner dynamics of the relationships of its elements with the environment" (Broens, 2004, p. 165, my translation).

The author also points out that, just as Debrun had posited, *it is through learning and adjustments in relation to the neighboring context that these innovations can be incorporated.* Therefore, an organism self-organizes (on a secondary level) to reformulate its modes of interaction with the environment. The subject is, thus, understood as able to learn: "The subject as defined by the theory of Self-Organization is every system able to learn, to incorporate innovations, i.e., to promote a secondary Self-Organization in the relations it maintains with the environment" (Broens, 2004, p. 166, my translation).

The members of both the SEEs discussed earlier have gotten together with a common purpose (that is, the need for employment and/or for improving their life conditions). At the time of their formation, corresponding to the SEE implementation phase, these SEEs can be understood as primary Self-Organizations, which, as Debrun characterizes them, correspond to "every process of integration of actually distinct elements that, instead of tending towards a given attractor, progressively consolidates its own attractor, therefore creating itself as a system" (Debrun, Chapter 2, this volume). Hence, each of these SEEs constituted itself as a system, in the sense that Bresciani Filho and D'Ottaviano (Chapter 3) understand the idea of a system as a set of elements having a structure and a function. The functioning of the system refers to the articulated set of activities of the elements of this system: "The functioning of a system is conferred by the set of activities of its elements, which conduct the process of transformation, exercising functions in a dynamic way but conditioned by the partial structure" (Chapter 3). For these authors, a system can be identified by its states, and the evolution of the system with the changes of these states; such changes of state can result from the behaviors of the elements of a system and from these elements' relationships with the environment.

I, therefore, suggest that the SEEs discussed in this chapter – having their own identities (defining their structures) and common objectives – have at present a certain functioning. My observations indicate that when making decisions, the SEE members need the aid of people from the GHS/IRCP and of other researchers; they cannot by themselves perform several types of work, such as elaborating projects, budget spreadsheets, etc. Moreover, they can only autonomously execute tasks of a more practical nature that can be defined as deriving from empirical knowledge.

Consequently, even though the SEEs have a certain degree of autonomy, the members do not have total autonomy in making some decisions and performing some actions, due to difficulties that arise in the execution of their tasks. This situation has been observed by myself and others in our time spent with the enterprises.

In relation to the women's carpentry shop (the MCF), for example, we noticed a situation in which the cutting angle of a piece of wood required some adjustments to be made to a miter saw. This was necessary in order to make a piece of wood that would fit into a window that was being produced. We noticed that the carpenters were only able to adjust the saw after the cutting angle had been given to them by a member of the supporting team. The carpenters were not able to calculate the measurement of the cutting angle by themselves, even after having observed and performed the adjustments several times. Because of this, we realized that they could only perform the manual part of the task and could not comprehend the required mathematical reasoning (Meneghetti and Daltoso Jr., 2009).

This was also reinforced in the interviews, as in Carpenter B's declaration when asked about difficulties related to the carpentry shop:

> for example, making a table, what do we know how to do? We can operate the machinery, yes, but we don't know how much wood is needed, what is needed, if it's only nails, if it's only glue, if it's nails and glue, *we don't know these kind of things yet, but we can learn, can't we? I believe so.*
>
> *(Carpenter B, MCF, my italics)*

Members F and G from the cleaning cooperative have also pointed out some of the difficulties met with in this SEE's daily routine: "Well, we do much of the budgeting, don't we? So, this is difficult for us. We use this square meter thing a lot and this is difficult [….]" [Member F]; "When we have to make, for example, an electronic bidding, it is complicated" [Member G].

We also have identified a dependency on the use of calculators in the case of the cleaning cooperative, due to difficulties in performing the basic mathematical operations required in the activities of this SEE: "We use the calculator a lot, everything we have to do related to mathematics we use the calculator. I already have some trouble with mathematics; without the calculator, I'm nothing" [Member F]; "They have a lot of difficulties, don't they? They use the calculator all the time. [...]" [Respondent B, member of the IRCP, accompanying this SEE].

In accord with Peirce (1958, cited by Andrade, 2011), I understand that the existence of doubt indicates the possibility of altering habits. For Peirce, doubt (as the antithesis of habit) paralyzes behavior. This opens up the possibility for a restructuring of the organization of the kind that characterizes secondary Self-Organization. It is under the influence of doubt that the subject-face strives for a reorganization of its habits.

As Andrade (2011, p. 89) emphasizes, it is possible to have secondary Self-Organization in at least two situations: "(i) in a reflection about our behavior and in the identification of the need for changes in at least part of this behavior, and (ii) in the establishment of a doubt over the efficiency of a habit" (my translation). In this sense, I understand that the SEEs have the potential to become secondary Self-Organizations, and I believe learning has a key role in this process. However, I must point out that, as Andrade (2011) affirms, Self-Organization as a result of the restructuring of an organization must originate from the spontaneous interaction (not imposed by a supervising instance) between an agent and his or her context of activity (physical, social, cultural).

It is also important to notice that, according to this author, it is to a certain extent the system/subject that decides or opts for the learning of a new technique or the acquiring of new knowledge. Therefore, when questions came up regarding the possibility of educational work in mathematics being done with the cleaning cooperative (in this case, a pedagogical workshop), I noticed a motivation, a desire to learn the concepts necessary for this SEE's activities: "I would really like to learn this math" [Member S]; "Hum… I hope the workshop can help us" [Member E]; "I would like to have the workshop, I really want to learn from it [...]"; [Member F]; "Look, if we learn a little math, it would help us" [Member G].

Openness and the willingness to learn were also present in the utterances of the carpenters, as seen in Carpenter B's previously quoted remarks. This evidences the need and the desire to move to a new level of operation, or of complexity, in Debrun's terminology. Learning is, thus, regarded as a means through which one can move from one organizational level to another one, the next level being superior in the sense of allowing for a greater autonomy for the group in its actions and decision-making.

In light of this, I understand that it is from the reflection of the subjects (SEE members) on their wishes and desires that the possibility arises for a restructuring of the system, or, in other words, for the occurrence of secondary Self-Organization. As SEEs, the groups must reach a new level of organization in regard to self-management, something that is aimed at in the context of solidarity economy. For this to occur, it is important to replace old habits with new ones, which can also happen through learning. Gonzalez et al. (2000, p. 75) affirm:

> the ability to create and change habits allows organisms to act in a way that favors their survival, and allows them to adjust their behavior according to the characteristics of the environment, modifying the environment and being affected by the modifications in accordance with the circular causality dynamics.
> *(my translation)*

By understanding their own functioning and by acquiring new knowledge – especially that related to mathematics – through processes of learning, these individuals can perform their activities with greater autonomy. This can lead to a change away from habits that are based in automated actions without reasoning and marked by the need for support from other people. When the individuals change, there will also be a change in the SEEs, for in a self-organized system, the whole influences the parts and the parts influence the whole.

Acknowledgments

I would like to thank FAPESP (Fundação de Amparo à Pesquisa do Estado de São Paulo) for financial support, as well as Prof. Dr. Ubiratan D'Ambrosio and Prof. Dr. Itala M. Loffredo D'Ottaviano, and all colleagues who contributed to the making of this research.

References

Andrade, R. S. C. (2011). *Sistêmica, hábitos e auto-organização*. PhD thesis, Instituto de Filosofia e Ciências Humanas IFCH/UNICAMP, Campinas, SP.

Brandão, C. R. (1986). *Educação Popular*. São Paulo: Brasiliense.

Brasil (2006a). Ministério do Trabalho e Emprego (MTE). Secretaria Nacional de Economia Solidária (SENAES). Atlas de Economia Solidária no Brasil. Brasília, DF: Author. Available at: www.mte.gov.br/ecosolidaria/sies_atlas_parte_1.pdf. Accessed on April 12, 2012.

——— (2006b). Ministério do Trabalho e Emprego. Secretaria Nacional de Economia Solidária (SENAES), Secretaria de Políticas Públicas e Emprego (SPPE), Departamento de Qualificação (DEQ). *I Oficina Nacional de Formação/Educação em Economia Solidária: documento final*. Brasília, DF: Author. Available at: http://cirandas.net/cfes-nacional/i-oficina-nacional-formacao-es.pdf. Accessed on April 12, 2012.

Broens, M. C. (2004). Sujeito e Auto-organização. In: Souza, G. M., D'Ottaviano, I. M. L., and Gonzalez, M. E. Q. (Orgs.), *Auto-Organização: estudos interdisciplinares* (pp. 159–176). Campinas: CLE/ UNICAMP. (Coleção CLE, v. 38).

FBES. (2006). *Forum Brasileiro de Economia Solidária: a experiência de gestão e organização do movimento de Economia Solidária no Brasil*. Available at: www.mobilizadores.org.br/wp-content/uploads/2014/05/texto-5363c725c2c79.pdf. Accessed on February 15, 2018.

Gaiger, L. I. (Org.). (2004). *Sentidos e Experiências da Economia Solidária no Brasil*. Porto Alegre: Editora da UFRGS.

Gonzalez, M. E. Q., Broens, M. C., and Serzedello, J. (2000). Auto-organização, Autonomia e Identidade Pessoal. In: D'Ottaviano, I. M. L., and Gonzalez, M. E. Q. (Orgs.), *Auto-Organização: estudos interdisciplinares* (pp. 69–81). Campinas: CLE/UNICAMP. (Coleção CLE, v. 30).

Icaza, A. M. S., and Tiriba, L. (2003). Economia Popular. In: Cattani, A. (Org.), *A Outra Economia*. Porto Alegre: Veraz Editores.

Kliksberg, B. (2002). *América Latina: uma região de risco – pobreza, desigualdade e institucionalidade Social*. Tradução de Norma Guimarães Azeredo. In: *Cadernos UNESCO Brasil. Série Desenvolvimento Social*. Brasília: UNESCO.

Kruppa, S. M. P. (2005). Uma outra economia pode acontecer na educação: para além da Teoria do Capital Humano. In: Kruppa, S. M. P. (Org.), *Economia Solidária e Educação de Jovens e Adultos*. Brasília: INEP (Instituto Nacional de Estudos e Pesquisas Educacionais Anísio Teixeira).

Lechat, N. M. P., and Barcelos, E. S. *Autogestão: desafios políticos e metodológicos na incubação de empreendimentos em Economia Solidária*. Available at: www.periodicos. ufsc.br/index.php/katalysis/article/view/5381/4736. Accessed on November 16, 2009.

Matsuda, P. M. (2008). Incubação de cooperativas populares e a extensão universitária – Estudo de caso na IRCP/UFSCar Incubadora Regional de Cooperativas Populares Universidade Federal de São Carlos. In 6 *Anais do IV Simpósio Acadêmico de Engenharia de Produção*. Viçosa-MG. Available at: www. saepro.ufv.br/Image/artigos/Artigo18. pdf. Accessed on November 16, 2009.

Meneghetti, R. C. G., and Daltoso Jr. S. L. (2009). A matemática utilizada por um grupo de marceneiras: um olhar inicial. In: 6 *Anais do Congresso Internacional de Educação: Educação e Tecnologia: sujeitos (des)conectados?* GT09. Educação Matemática e processos de (in)exclusão escolar. (pp. 374–383). UNISINOS-São Leopoldo. RS. Cd-Rom.

Peirce, C. S. (1958). *Collected Papers of Charles S. Peirce*. Cambridge, MA: The Harvard University Press.

Singer, P. (2002). *Introdução à Economia Solidária*. São Paulo: Editora Fundação Perseu Abramo.

Singer, P., and Souza, A. R. (2000). *A Economia Solidária do Brasil – A autogestão como resposta ao desemprego*. São Paulo: Contexto.

Veiga, S. M., and Fonseca, I. (2002). *Cooperativismo – uma revolução pacífica em ação*. Rio de Janeiro: DP&A/Fase.

12

MULTILINGUALISM AND SOCIAL SELF-ORGANIZATION IN BRAZIL

Claudia Wanderley

Introduction

Multilingual reality in Brazil includes public verbal silence and public verbal censure as part of the historical constitution of multilingualism. Mother tongues whose verbal expression was publicly censored verbally entered into a Self-Organization process so as to continue signifying in non-logocentric ways in public space. In this chapter, we recognize three main instances in Brazilian history in which languages were censured and silenced by forbidding the realization of multilingualism in our territory and propose the initial elements of a hypothesis for understanding the response to the censure of mother tongues in Brazil as a primary Self-Organization process.

A multilingual Portuguese-speaking country

The most important multilingual background elements in the Brazilian sociolinguistic situation are Amerindian languages, African languages, and, later, the languages of communities of formal immigrants.

Rodrigues (2002) estimates that the number of existing languages of the original peoples living in our territory in the sixteenth century was 340, and that the population at that time was 6 million people. In 2010, the self-declared indigenous population of Brazil was 817,963, and an estimated 300 indigenous languages were spoken[1] (Figures 12.1 and 12.2).

A plurality of languages came to Brazil from the African continent. We estimate that four million people arrived on our shores, altogether speaking approximately 100 African languages; Alencastro mentions the existence of around eighty languages in one South American port: "The Ignatian writer Alonso de Sandoval, in his research carried out on the ships anchored in Cartagena in the

FIGURE 12.1 Two Tupinambá Chiefs, Hans Staden (1525–1576), Português: Xilogravura. Dois Chefes Tupinambás com os Corpos Adornados por Plumas – ilustração do livro "Duas Viagens ao Brasil" de Hans Staden (1557), Wikimedia Commons.

FIGURE 12.2 Tapuia woman (1641). Albert Ekhout (1641–1644), Wikimedia Commons.

early seventeenth century, recorded more than seventy languages and dialects among the deportees" (Alencastro, 2000, p. 148).

After the end of slavery in 1888, the Brazilian government instituted a policy of inviting immigrants to work on the land, and these individuals also contributed to the rich linguistic and cultural situation in Brazil with the arrival of approximately eighty immigrant languages (in another study, based on an open estimate, I counted eighty immigrant languages in the São Paulo region alone; see Wanderley, 2009).

Amerindians, an estimated 340 languages

The majority of indigenous languages in the country are in danger of disappearing, having very few speakers. In the case of the indigenous population, two main factors played a strong role in the overall process of the encounter with Europeans. First of all, they were in their homeland, so getting away from the newcomers was easier. Second, the Portuguese crown could not control commerce inside the territory. The control of the Africans brought in conditions slavery occurred in the ports. Because the capital of the Portuguese empire in Lisbon could not control the commerce of original peoples, it discouraged their enslavement and encouraged Jesuit participation in their "salvation".

The promotion of the erasure of indigenous habits and ways of living is very well documented, and was seen as a kind of evaluation parameter that signified the success of the colonial enterprise. In 1758, the Marques de Pombal affirmed:

> It has always been the maxim unalterably practiced in all nations which have conquered new domains, to soon introduce their own language to the conquered peoples, for it is indisputable that this is one of the most effective means of banishing from rustic peoples the barbarity of their old customs.[2]

The effort to promote the erasure of the original peoples' spoken languages, and to force the consolidation of a logocentric monolingual world, is yet to have its effects well understood. Ilari (2006, p. 64) points to testimony that indigenous languages were still spoken by a considerable part of the population in 1822 (cf. Lobo 2001, p. 164), when Brazil became independent from Portugal. In indigenous villages,[3] it is understood that hearing is a strong ability that one possesses when present in the forest environment. If we add censure of one's mother tongue in social and public life to these already strongly developed skills of reception of information, we can see that there was a good chance that new complex layers of reception of information developed as an effect of the silence first imposed in the colonial period (1500–1822) by the Portuguese empire, and afterward by the independent nation of Brazil.

Human trafficking and an estimated 100 languages

The exploitation of our workforce was basically done by Portugal in partnership with the Catholic Church. The Church would take care of souls by means a small tax paid by Portuguese Crown for baptism (at the port on leaving Africa) and christening (at the port on arriving in Brazil). Human traffic promoted by Portugal, allied with conversion to Christianity, became a fundamental source of wealth in this period (1551–1860).

Portuguese *negreiros* (slave ships) circulated widely on the Ibero-American market in the sixteenth century, and were responsible for the importation of approximately four million people into Brazil until the nineteenth century (Curtin, cited in Alencastro, 2000, p. 184). In this context, goods and slaves were exchangeable; they were synonyms. The emphasis in our reading of Alencastro is on the process of de-socialization (adopted from Claude Meillassoux):

> Slavery has two processes: the first is depersonalization, and the second is a de-socialization; that is, a stranger is taken from his community, his country, his nation, his language, and his religion, is taken elsewhere. The slave is always a stranger. And, in this other place, he becomes a thing, is depersonalized. He becomes merchandise, cattle, the moment he is chained. Iron is the mark of the tax paid to the Crown.[4]

The goal was to make profit with human traffic from Africa, and with a workforce in the production of goods in South America. The Portuguese also created symbolic conditions for four million people to be deported from Africa to Brazil, and to have their histories erased. The promotion of de-socialization and the process of active forgetfulness is also supported by linguistic strategies; as a linguist thinking in a postcolonial critical perspective (Spivak, 1999; 2006), I would enumerate the following: (1) silence,[5] (2) censure,[6] and (3) *forclusion*.[7]

According to our hypothesis, the language policy consistently applied in Brazil is "verbal silence" in public space. And due to this impossibility to say what one wants to say, it is necessary to resist and express oneself outside of the verbal domain (Figure 12.3).

Immigrants and an estimated eighty languages

After the end of slavery, nineteenth century, the Brazilian government had a policy to invite immigrants to work as a colonist. These groups also contributed to the rich linguistic and cultural situation in Brazil. Policies of nationalization [or suppression of linguistic plurality] silenced the newly arrived in Getúlio Vargas period.

In 1938, Vargas launched a nationalization campaign in Brazil. Languages, which were not Portuguese, were censured in order to maintain Portuguese

FIGURE 12.3 Jean Baptiste Debret in Brazil, XVIII century, portraits the market at the Valongo street (Mercado da rua do Valongo), Rio de Janeiro, Brazil. Wikimedia Commons.

as the dominant language in the territory. The language of schooling and public space in Brazil was once again defined to be Portuguese. The policies went even further: only Brazilians could be school teachers, all the classes were to be in Portuguese, and foreign language classes were forbidden for anyone under the age of fourteen. In 1939 in Brazil, it became forbidden to speak foreign languages in public, including in religious ceremonies, and the Brazilian Army was delegated to supervise monolingualism in immigrant communities.

The temporary results of the Vargas campaign of nationalization were that the Army was inside the schools and that a great movement of resistance against monolingualism arose in immigrant communities. It is easier to identify the origin of these immigrants, and their mother tongues, for they were treated as citizens and documented their arrival at port. On this level, we have an institutional written narrative for this group.

With Vargas nationalization campaign, new generations gradually left off their parents' and grandparents' mother tongues and became monolingual. In this case, there is a different kind of organized memory of linguistic erasure from that of Afro-Brazilians and Amerindians. There is a memory allowed if this memory is in national language. It is a type of memory already established in the domain of the logocentric world of the nation-state.

Mother tongue censure and Self-Organization

The presupposed contingent of around 520 languages constitutes the linguistic heritage of Brazil.[8] The hypothesis that we present here is that as mother tongues were publicly censored in verbal expression, the meanings they carried went through a Self-Organization process in order to keep signifying. This is a first estimate of the three main linguistic currents (Amerindian languages, African languages, Immigration languages) that were censured and silenced but found ways to promote senses in non-logocentric ways.

With regard to primary Self-Organization, Debrun (Chapter 2, this volume) states: "every process of integration of actually distinct elements that, instead of tending towards a given attractor, progressively consolidates its own attractor, therefore, creating itself as a system".

In our hypothesis, the estimated 520 languages of the linguistic heritage in Brazil are historically consolidated in layers of meaning outside of verbal expression. These languages are making sense in our present context of out-of-control logocentrism. The hypothesis is that considering these phenomena as a primary Self-Organization process, this historical process of silencing mother tongues in public spaces was converted into a complex non-verbal system of expressions. For Debrun:

> In "primary" Self-Organization, identity does not exist in the starting point. [...] What happens is just that the identity, as it develops, leads to what Spinoza thought of as the tendency of a substance to remain within its substance, which means that an immanent finality emerges "adhered" to the being, in the sense that, although not pursuing any goal or target, the being "adheres" to its own existence.
>
> *(Debrun, Chapter 2, this volume)*

There is a kind of linguistic identity developed in the process of making meaningful statements outside the domain of public verbal expressions, possibly a more complex developed capacity of receiving information. Each instantiation demands a proper study of the elements, attractors, and dynamics of the system, for the conditions of the linguistic heritage and the elements used to promote the censorship are different. There is no evidence that there are similar dynamics at work in each case. Debrun says about the process of primary Self-Organization that: "This means that the attractor of the process is the process itself, which tends to 'attach to itself', or even to crystallize. The process, as we have seen, is self-referent in a certain way" (Debrun, Chapter 2, this volume).

The strong symbolic and physical efforts of the colonizers to consistently promote erasure of people's mother tongues probably also promoted specialized dynamics of the capacity of hearing, reading, analyzing, and capturing senses and meanings on a broader level, which might point to a differentiated process of the reception of information.

I propose that there are many systems of self-reference forged in historically acquired silence outside of the logocentric public dominion, and that these should be considered part of our multilingual constitution. In the case of Brazil, this concerns speakers of non-national languages and the meaningful resistance to censorship, the possibility of symbolic survival, and layers of complex information kept and expressed along with public verbal silence. To study these publicly censured heritages, it is important to ethically express the suspension of this historical censorship.

Notes

1 IBGE – Instituto Brasileiro de Geografia e Estatística. Available at: https://indigenas.ibge.gov.br/graficos-e-tabelas-2.html.
2 Our translation; see: http://lemad.fflch.usp.br/sites/lemad.fflch.usp.br/files/Diret%C3%B3rio%20dos%20%C3%8Dndios%201755b.pdf.
3 In this respect I am very thankful to the Paiter Suruí People, especially, Chief Almir Suruí, Professor Arildo Suruí, and Professor Rubens Suruí, for the opportunity of regularly being their guest at Sete de Setembro Indigenous Territory since 2016.
4 Our translation; see https://obenedito.com.br/corpo-na-america-e-alma-na-africa/.
5 For that notion of silence, see Ducrot (1991), and the "geology of silence" in Bachelard (1943, p. 174):

> If we want to study this integration of silence with the poem, we must not make it the simple linear dialectic of pauses and splinters along a recitation. It must be understood that the principle of silence in poetry is a hidden thought, a secret thought. As soon as a thought able to hide under its images watches a reader in the dark, the noises choke, the reading begins, the slow pensive reading. In search of a hidden thought under expressive sediments the geology of silence develops.
>
> (our translation)

6 For the notion of censure, see Salazar (2004, p. 12): "Censure is simply the expansion of the speech-act".
7 For the notion of *forclusion*, see Gayatri Chakravorty Spivak (1999, p. 4):

> I borrow the term "foreclosure" (*forclusion*) from Lacanian psycho-analysis. I read psychoanalysis as a technique for reading the pre-emergence (Raymond Williams's term) of narrative as ethical instantiation. Let me sketch this technique briefly by way of the entry for 'Foreclo-sure' in a still useful general lexicon of the passage between Freud and Lacan, *The Language of Psycho-Analysis*.

8 We are not going to address the Portuguese language in this work; there are already many works addressing the Brazilian national language; for instance, Rodolfo Ilari in "O Português da Gente: a língua que estudamos e a língua que falamos"; São Paulo, Ed. Contexto, 2006. It should also be remembered that Portuguese-speaking people were a small percentage of local population in the sixteenth century.

References

Alencastro, L. F. (2000). *O Trato dos Viventes: formação do Brasil no Atlântico Sul*. São Paulo: Companhia das Letras. p. 524.
Bachelard, G. (1943). *L'air et les songes. Essai sur l'imagination du movement*. Available at: http://classiques.uqac.ca/classiques/bachelard_gaston/air_et_les_songes/air_et_les_songes.pdf.
Ducrot, O. (1991). *Dire et ne pas dire Principes de sémantique linguistique*. Paris: Hermann.
Ilari, R. (2006). *O Português da Gente: a língua que estudamos e a língua que falamos*. São Paulo: Contexto.

Lobo, T. (2001). Cartas Baianas Setecentistas. São Paulo: Humanitas. pp. 164–165.

Rodrigues, A. D. (2002). *Línguas Brasileiras: para o conhecimento das línguas indígenas*. São Paulo: Edições Loyola. 4a edição Junho de 2002.

Salazar, P.-J. (2004). Censorship: a philological (and rhetorical) viewpoint. *The Public*, 11(2), pp. 5–18.

Spivak, G. C. (1999). *A Critique of Postcolonial Reason: Toward a History of the Vanishing Present*. Cambridge, MA: Harvard University Press, p. 449.

———. (2006). *In Other Worlds*. New York: Routledge, p. 409.

Wanderley, C. (2009). A periferia digital e o processo de descolonização. In: Galves, C., Garmes, H., and Ribeiro, F. R. (Org.), *Africa-Brasil Caminhos da Lingua Portuguesa*. Campinas: Unicamp, p. 225–243.

PART IV
Semiotic studies

13

PROLEGOMENA TO A SEMEIOTIC THEORY OF SELF-ORGANIZATION

Vinicius Romanini

Introduction

It is well known that the theory of signs envisioned by the North American philosopher and logician Charles Sanders Peirce (1839–1914) is part of a wider philosophical architecture. Peirce's system is structured from mathematics, but also involves metaphysics, phenomenology, and a set of normative sciences composed of the triad of ethics, esthetics, and logic. The latter, especially important for the purposes of this chapter, is understood as the doctrine that studies the general conditions for the production of the representation and sharing of meaning in a community of inquirers interested in the pursuit of truth – a process which is also known as semeiosis.

Semeiotic,[1] for Peirce, is a synonym for logic taken in its wider sense, and so comprises a general theory of information and communication. We will argue here that a semeiotic theory of Self-Organization must take into account how different types of signs relate to one another to produce what we understand to be self-organizing entities. As Kant taught us, our conceptions of external reality are transcendental creations of our own mind, and we will not be able to understand Self-Organization if we do not, first, understand how our minds idealize it.

Because the task is huge and complex, this chapter will deal mainly with that part of Peircean logic which focuses on speculative grammar, or the study and classification of all possible types of signs taking part in the process of the representation and interpretation of whatever comes into our minds, either from perception or imagination. But why begin the study of a semeiotic theory of Self-Organization with Peirce's classification of the sign? First, because the Peircean Semeiotic, as the logic of semeiosis – the action of signs as they produce intelligible effects – can be the basis for a systemic and evolutionary (that is, dynamic and complex) theory of reality; second, because scientific classifications are important for disclosing the logical determinations ruling the phenomena apprehended by experience.[2] Since self-organizing systems must be identified as such by a community of inquirers in order to become the object of scientific research, being able to pinpoint, classify, and relate the classes of signs cooperating in the formation of our conceptions of Self-Organization is the first and most important step toward a sound theory.

Classifications are not definitive, but even when arranged in a hypothetical way and subject to continuing improvements, classifications provide a precious diagram of our knowledge of a portion of reality. On this basis, we as a community of researchers can retroductively[3] extract information that was not immediately visible before such classifications were disclosed. When newfound information passes through the sieve of experience, new knowledge is consolidated, which then becomes available to the members of the community.

Here, we have the first glimpse of what sort of epistemology is involved when one tries to build a theory of Self-Organization from semeiotic principles: self-organizing systems are those capable of development and differentiation, which implies a classificatory gradient from the simplest to the more complex. If this is indeed the case, self-organizing systems must be capable of the perception, representation, and interpretation of certain final causes. In other words, they must be semeiotic in their nature.

Following this line of thought, I will present here an explanatory model for the structure of semeiosis that I call the *Solenoid of Semeiosis*. It is based on minute distinctions of aspects of the sign and its categorical value. We believe that such a model can both help us find the complete classification of all sixty-six classes of signs envisioned by Peirce, as well as show the evolutionary dynamism of semeiosis, opening the doors to a semiotic theory of reality that includes the study of Self-Organization processes. Once its grammatical roots are established, Self-Organization can then be the object of study of general rhetoric or methodeutic, which is interested in discovering how certain arguments allow us to understand and connect parts of the real that otherwise would be left unintelligible.

From phenomenology to semeiotic

Peirce considers phenomenology the science that intends to identify the fundamental constitutive elements of any phenomena present in our minds, regardless of their nature or ultimate reality. The starting point of phenomenology is precisely the *phaneron* (a Greek word meaning what is manifest, or apparent, to our minds). Peirce defined it as the unanalyzed set of everything that is immediately present: qualities of feelings, percepts, cognitions, memories, expectations, etc.[4]

It is due to this etymology that in some of his mature writings, Peirce preferred the term "phaneroscopy", meaning something that "describes the apparent". As all knowledge enters through the doors of perception, understanding the phenomenology of perception is the first and most important step in assuring a strong epistemology. Following the philosophical tradition of Aristotle and Kant,

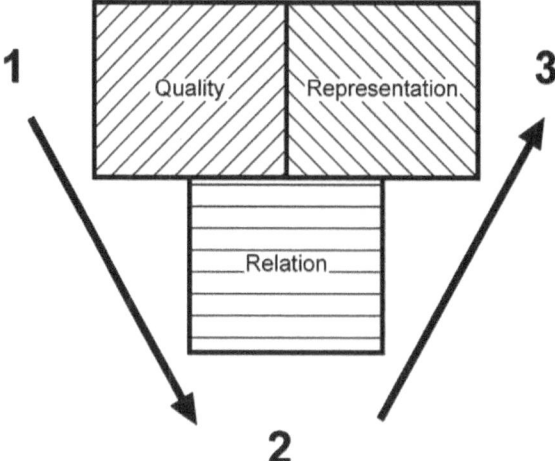

FIGURE 13.1   Peirce's three categories as originally presented. "Relation" as later changed for "reaction".

who built their systems on tables of categories, Peirce elaborated a list of three fundamental predicamenta: quality, relation, and representation, that he subsequently renamed as firstness, secondness, and thirdness. He then set them in the order of classification shown in Figure 13.1.

Firstness is pure possibility, originality, spontaneity, feeling, and pure quality; something that is in itself. Secondness is pure existence, reaction, shock, singularity, and alterity; something that is only related to a second thing. Thirdness is the continuous, the habitual, process, pattern, mediation, representation; something that is the synthesis of relationships.

Although none of those three categories is reducible to any of the remaining ones, we must not expect to find them in a pure state in the phenomena. In truth, the phaneron is always composed of combinatory gradients among firstness, secondness, and thirdness, appearing in our minds in a dynamic and continuing way, and producing a kind of "screen" or phenomenological "topos" in constant transformation.

Semeiotic and pragmatism

As the logic of pragmatism, semeiotic is the doctrine of signs that are able to grow and develop with experience, while at the same time producing effects that can change, as in a feedback loop, the environment where they occur. Semeiotic has three branches: pure grammar, which tries to describe the fundamental aspects and relationships of each sign, so as to find the possible types of signs and make possible their classification; critical logic, which studies the conditions that allow the sign to represent its object; and methodeutic or rhetoric, which is essentially a theory of the communication. This last branch of semeiotic studies shows how each sign communicates the form of its object in order to create logical interpretants such as propositions, inferences, and arguments.

Peirce was indeed a pioneer in extending traditional logic to general semeiotic. He developed his ideas until he reached the logical diagrams that he called existential graphs, a system that was able to include both propositional logic and the calculus of predicates. Nevertheless, he left his semeiotic incomplete. We believe that the main reason for this was the difficulty he had in finding and describing all aspects of the sign in a way that was sufficient to produce a classification that included both variable time dynamics and modality in an evolutionary logic where possibilities are updated while guided by teleological signification processes.

The analytical cascade of the phaneron

In a constructive strategy for discovering all the aspects of the sign, from the simplest ones to the most complex, Peirce identified ten of these aspects. In a previous work (Romanini, 2006), we adopted the opposite strategy, choosing to start from the perfect sign, formed by the threefold relationship of sign (S), dynamic object (DO), and final interpretant (FI), represented as S-DO-FI. From this starting point, we then extract analytically all of the aspects involved, as well as their possible relationships.

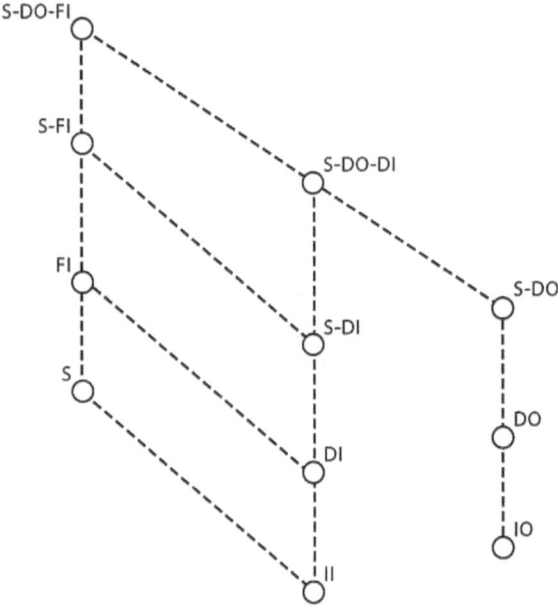

FIGURE 13.2 The analytical cascade of the phaneron places the most complex sign aspect on top and then proceeds to the analysis creating three axis that shows eleven sign aspects in total.

We call this analysis the analytical cascade of phaneron (Figure 13.2), which naturally produces eleven aspects, as opposed to Peirce's ten. The additional aspect that is not part of Peirce's analysis is the relationship between the sign itself, the dynamic object, and the dynamic interpretant (S-DO-DI).

Descriptions of the aspects of the sign

We now offer a short description of each of the eleven aspects of the sign as we understand them:

a   Immediate object (IO): This is what is immediately present in the phaneron, forming the ground or substrate of the sign itself. It is the aspect of the dynamic object selected by the sign to represent it.

b   Immediate interpretant (II): This is any effect immediately produced by the sign as it incorporates the immediate object.

c   Dynamic object (DO): This is anything that can be represented by a sign (a feeling, an emotion, a quality, an idea, an argument, a book, the universe). The dynamic object is always outside the sign. However, the sign is put in such relationship with the dynamic object as to produce an interpretant (effect) somehow compatible with the very effect that the dynamic object itself would produce in the mind of an interpreter. Since the sign is always incomplete in relation to its object, a sign can only select some aspect of a set of aspects of the dynamic object, which is its immediate object.

d   Dynamic interpretant (DI): It is the effect effectively produced by the sign itself, regardless of what it represents.

e   Final interpretant (FI): It is the destined, telic, or intending effect that would be produced by the sign itself at the end of semeiosis, regardless of what it represents.

f   Relationship between sign and dynamic object (S-DO): This is the way the sign is related to the object that it professes to represent (its dynamic object). It can be similar to its qualities, it can be physically connected to it, or it can be related to it by some kind of habit (either conventional or natural).

g   Relationship between the sign and the dynamic interpretant (S-DI): This is the way the sign relates to its dynamic interpretant, eventually consolidating its ability to influence and produce concrete effects, be they physical or mental.

h   Relationship between the sign and final interpretant (S-FI): This is the way the sign relates to the final interpretant in order to develop towards a virtual purpose.

i   Relationship between the sign, dynamic object, and dynamic interpretant (S-DO-DI): This threefold relationship describes the ability that a sign has to incorporate information from the dynamic object and communicate it to an interpreting mind, so as to produce concrete effects within the scope of the sign.

j   Relationship between sign, dynamic object, and final interpretant (S–DO–FI): This threefold relationship describes the ability a sign has to create self-organizing semeiosis in order to communicate to its interpreting minds the information gathered from the dynamic object and thus achieve a virtual purpose.

Also, note that the analysis of the perfect sign allows for identifying three different axes that together operate as a structural frame for the semeiosis:

1   Objectivation, formed by the sequence:
    S–DO–FI ---- S–DO–DI ---- S–DO ---- DO ---- IO
       This is the axis that points to the dynamic and immediate objects of the sign, its internal and external propellants.
2   Interpretation, formed by the sequence:
    S–DO–FI ---- S–DO–DI ---- S–DI ---- DI ---- II
       This is the axis that points to the dynamic and immediate objects of the sign, its internal and external sensible effects.
3   Significance, formed by the sequence:
    S–DO–FI ---- S–FI --- FI ---- S
       This is the axis that points to the aspects that describe the teleological growth of signs.

Each aspect of the sign must be subsequently analyzed according to the dominant category that characterizes it, that is, it must be trichotomized in the three phenomenological categories. In what follows, we offer our description of each aspect of the sign, as well as the meaning it assumes, in terms of the threefold division into the categories.

   Note that the phaneroscopic analysis of the sign allows us to penetrate into the deeper relationships between phenomenology and semeiotic. Even more importantly, if we accept that the ontology of reality has the same structure as that shown by phenomenology – that is, if we accept the Peircean hypothesis that the universe is full of signs if not really composed solely of them – the aspects identified in the axes of the analytical cascade are also those aspects that compose reality itself.

The solenoid of semeiosis

So far, what we have done is to isolate the sign, outlining it as much as possible and then decomposing it into its basic elements and the relationships among those same elements. However, we know that the perfect sign is essentially dynamic, producing semeiosis. No matter how creative the semeiosis may be, it is never completely random, for absolute chaos is identical to nothing. The basic element of regularity in semeiosis is the flow of determination of categories among the aspects that bind the three axes described earlier.

That flow enables the natural ordering of the eleven aspects according to a sequence of four hierarchically organized bonds. Starting with the immediate object (IO), the order of determination of the different aspects may be organized as follows:

$$IO \geq II \geq S \geq DO \geq DI \geq FI \geq S\text{-}DO \geq S\text{-}ID \geq S\text{-}FI \geq S\text{-}DO\text{-}DI \geq S\text{-}DO\text{-}FI$$

This order is such that $1 \leq IO \leq 3$, where 1 means firstness, 2 means secondness, and 3 means thirdness.

For each aspect, its predecessor (if any) must have at least the same categorical value, and its successor (if any) must have at most the same categorical value. So, if II has a value of 3, IO must be 3 as well, and S can be either 1, 2, or 3. If II has a value of value 1, then IO can be 1, 2, or 3, and S can be only 1. The flowing of possible information produces a solenoid form which we call the *Solenoid of Semeiosis*.

Semeiosis and periodicity

Semeiosis is presented as a periodic flowing. By periodicity, we mean the phenomenon of repetition of a set of properties at regular intervals (Scerri, 1998), although there is an increase of complexity in the whole. The four periods of the Solenoid of Semeiosis are shown in Figure 13.3.

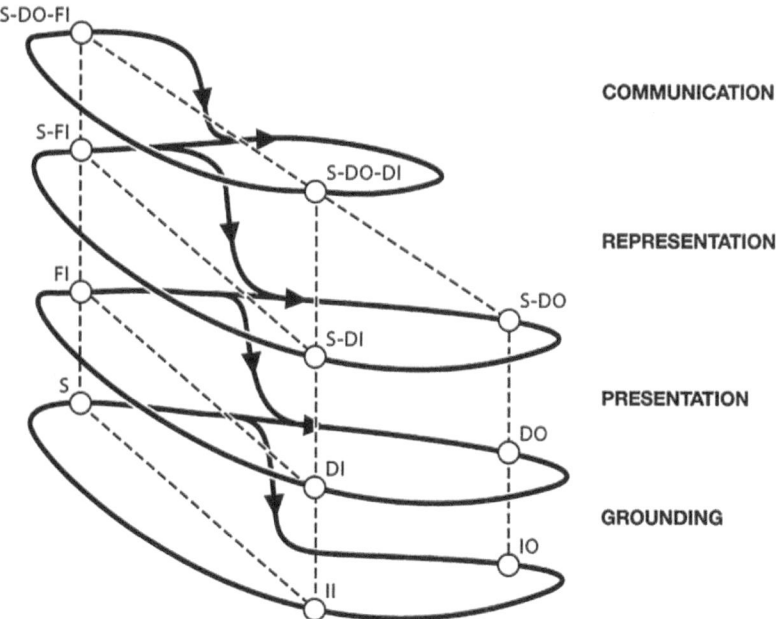

FIGURE 13.3 The Solenoid of Semeiosis is a diagram that connects the eleven sign aspects according to their order of determination, starting from the immediate object (IO) and ending at triadic relation among sign, dynamic object and final interpretant.

Grounding

During this period, there occurs what Peirce calls the "habit of breaking habits",[5] or the trend toward breaking symmetry, which occurs as long as novelties are introduced into the phaneron. Here, space and time sensations emerge from the percepts and percipua to produce the basic patterns of the phaneron. While space is topologically presented, a natural trend toward homogeneity rules the process, producing entropy and the sensation of time passing.

Presentation

In this period, signs are presented as stimuli (irritations) creating the experience of otherness. The continuity of those stimuli produces patterns that are eventually developed into habits of behavior, and into systems ruled by information (the signs per se). When they show regularities and the property of permanence, they can be investigated as real.

Representation

In this period, the sign develops its ability to represent its dynamic object by internalizing information about it while it grows and acquires the power to transform reality, where it is manifested as real.

Communication

In this period, the sign, dynamic object, and interpretant are merged in a genuine threefold relationship able to produce communicative effects. Communication grows from the mechanical transmission of signals to the real sharing of meanings in a community of interpretants.

Semeiosis and autopoiesis

Semeiosis is autopoietic (Maturana and Varela, 1980, p. 78), which means that it produces itself from a fundamental complementarity between structure and function.

Semeiosis and development

Semeiosis is ampliative, beginning from the simple towards the varied and complex, that is, it goes in the direction of the increase of information.

Semeiotic information

We are not going to discuss Peirce's theory of information here. For the purposes of this article, it is enough to understand that for Peirce, information is a logical quantity tied to the growth of signs while they synthesize new qualities or increase their ability to represent objects. As Peirce explained in 1893: "Analog to the increase of information within us, there is a phenomenon of the nature – development – through which a multitude of things acquire a multitude of features, which were involved in few features of few things" (CP 2.434; Peirce 1958 will be henceforth referred to as CP, the number indicating volume and paragraph number, respectively).

In our solenoid of semeiosis, information has four phases, each one aligned to the other, corresponding precisely to the phases of the research through which we acquire knowledge and develop sciences. In each of these phases, we have the production of habits arising, being set, and eventually crystallized, interrupting the information process and staunching the semeiosis. This means that in each phase, information only occurs whenever there are habits being formed and in continuing transformation. One can even have information occurring simultaneously in two or more phases.

Furthermore, information has the ability to feed back into its own development process, a property that in the solenoid of semeiosis is represented by the four arrows returning from the meaning axis, closing the periods and creating hierarchies. The four informative phases are given in the following sections.

Perceptive

In this phase, habits are produced during the perception process. Space and time sensations emerge from the synthesis of percepts, and perceptive judgments from the percipua. Representamina and dynamic objects are entangled. Therefore, there is no possibility of information and knowledge. All the processes are non-conscious, and there is no sense of identity.

Inquisitive

In this phase, habits are produced by a continuing distinction between the representamina and its neighborhood, which then become possible dynamic objects of the signs. By means of the investigation process, the appearances of the phaneron acquire generality, producing information and knowledge. Signs are then animated by purposes, and they grow while their dynamic objects become the pattern of the reality.

Deliberative

In this phase, a habit is formed through participation in the regularities of the real, so as to produce living and growing symbols. Behaviors and practices are coded through a multitude of types of languages. Communicative actions are used to build a community of interpretants and to transform reality, producing culture.

Scientific

In this phase, habit is formed by means of communication inside a community of interpretants. The usual relationship (conventional or spontaneous in all of its gradient of possibilities) allows the creation of symbolic communication and the search for common purposes. Meanings are expressed in the form of scientific arguments, cultural habits, or, if the habit is crystallized, in the form of dogmatic communications and fundamentalist habits.

The razor of causality

The trichotomization of the eleven elements of the sign would allow, in the thesis, the production of seventy-eight arrangements. Nevertheless, not all of them are logically possible. Twelve of them are logical aberrations because they do not respect what we will call the *rule of the razor of causality*. Such a rule is necessary in order to preserve the reality of the secondness in semeiotic and, therefore, the casualty in the structure of the phaneron. The rule can be outlined as follows:

> Given the order of determination of the aspects in the Solenoid of Semeiosis, there should be an **n** amount of secondness occurrences in the axis of signification equal to the same amount **n** in each of the other two axes.[6]

For example, if the aspect of the representamen or sign in itself (S) has a value 2 (secondness), it must be existentiality connected either to the immediate object (IO) or to the dynamic object (DO), or both. The reason for this is that there cannot be a sign such as a fingerprint, for instance, without the representamen being existentially connected to a finger, which is its dynamic object (or, at least, the representamen must be connected to the percept and to the perceptible fact that brings to the mind the intrinsic qualities of that same fingerprint, which would be its immediate object and immediate interpretant).

After eliminating the twelve aberrant classes of signs, the total amount of logically possible classes is sixty-six (see Figures 13.4 and 13.5).

The sixty-six classes of signs can be arranged in a triangular figure (Figure 13.6) along the lines of the triangle of the ten genuine classes of signs presented by Peirce in 1903. (There are relationships between the classification adopted by Peirce, based on divisions of the three correlates, and ours, based on a division of eleven aspects. This subject will not be treated here.)

Semeiotic Matryoshkas

Periods and phases are hierarchically organized in a way that resembles the traditional Russian Matryoshka dolls. Another analogy would be to imagine developing vortices nested one inside the other. Thus, the grounding period is nested in the presentation period that hides it in its structure. The period of representation involves the two

CLASSES	TRICHOTOMIES IN THE ASPECTS										
	IO	II	S	DO	DI	FI	S-DO	S-DI	S-FI	S-DO-DI	S-DO-FI
1	1	1	1	1	1	1	1	1	1	1	1
2	2	1	1	1	1	1	1	1	1	1	1
3	2	2	1	1	1	1	1	1	1	1	1
4	3	1	1	1	1	1	1	1	1	1	1
5	3	2	1	1	1	1	1	1	1	1	1
6	3	3	1	1	1	1	1	1	1	1	1
7	2	2	2	1	1	1	1	1	1	1	1
8	2	2	2	2	1	1	1	1	1	1	1
9	2	2	2	2	2	1	1	1	1	1	1
10	3	2	2	2	1	1	1	1	1	1	1
11	3	2	2	2	2	1	1	1	1	1	1
12	3	3	2	2	2	1	1	1	1	1	1
13	2	2	2	2	2	2	1	1	1	1	1
14	2	2	2	2	2	2	2	1	1	1	1
15	2	2	2	2	2	2	2	2	1	1	1
16	3	2	2	2	2	2	2	1	1	1	1
17	3	2	2	2	2	2	2	2	1	1	1
18	3	3	2	2	2	2	2	2	1	1	1
19	2	2	2	2	2	2	2	2	2	1	1
20	2	2	2	2	2	2	2	2	2	2	1
21	2	2	2	2	2	2	2	2	2	2	2
22	3	2	2	2	2	2	2	2	2	1	1
23	3	2	2	2	2	2	2	2	2	2	1
24	3	2	2	2	2	2	2	2	2	2	2
25	3	3	2	2	2	2	2	2	2	2	1
26	3	3	2	2	2	2	2	2	2	2	2
27	3	3	3	1	1	1	1	1	1	1	1
28	3	3	3	2	1	1	1	1	1	1	1
29	3	3	3	2	2	1	1	1	1	1	1
30	3	3	3	3	1	1	1	1	1	1	1
31	3	3	3	3	2	1	1	1	1	1	1
32	3	3	3	3	3	1	1	1	1	1	1
33	3	3	3	2	2	2	1	1	1	1	1

FIGURE 13.4 Applying the razor of causality to the seventy-eight classes of signs generated by the eleven sign aspects, we end up with only sixty-six possible classes of signs.

CLASSES	IO	II	S	DO	DI	FI	S-DO	S-DI	S-FI	S-DO-DI	S-DO-FI
34	3	3	3	2	2	2	2	1	1	1	1
35	3	3	3	2	2	2	2	2	1	1	1
36	3	3	3	3	2	2	2	1	1	1	1
37	3	3	3	3	2	2	2	2	1	1	1
38	3	3	3	3	3	2	2	2	1	1	1
39	3	3	3	2	2	2	2	2	2	1	1
40	3	3	3	2	2	2	2	2	2	2	1
41	3	3	3	2	2	2	2	2	2	2	2
42	3	3	3	3	2	2	2	2	2	1	1
43	3	3	3	3	2	2	2	2	2	2	1
44	3	3	3	3	2	2	2	2	2	2	2
45	3	3	3	3	3	2	2	2	2	2	1
46	3	3	3	3	3	2	2	2	2	2	2
47	3	3	3	3	3	3	1	1	1	1	1
48	3	3	3	3	3	3	2	1	1	1	1
49	3	3	3	3	3	3	2	2	1	1	1
50	3	3	3	3	3	3	3	1	1	1	1
51	3	3	3	3	3	3	3	2	1	1	1
52	3	3	3	3	3	3	3	3	1	1	1
53	3	3	3	3	3	3	2	2	2	1	1
54	3	3	3	3	3	3	2	2	2	2	1
55	3	3	3	3	3	3	2	2	2	2	2
56	3	3	3	3	3	3	3	2	2	1	1
57	3	3	3	3	3	3	3	2	2	2	1
58	3	3	3	3	3	3	3	2	2	2	2
59	3	3	3	3	3	3	3	3	2	2	1
60	3	3	3	3	3	3	3	3	2	2	2
61	3	3	3	3	3	3	3	3	3	1	1
62	3	3	3	3	3	3	3	3	3	2	1
63	3	3	3	3	3	3	3	3	3	2	2
64	3	3	3	3	3	3	3	3	3	3	1
65	3	3	3	3	3	3	3	3	3	3	2
66	3	3	3	3	3	3	3	3	3	3	3

FIGURE 13.5 Applying the razor of causality to the seventy-eight classes of signs generated by the eleven sign aspects, we end up with only sixty-six possible classes of signs.

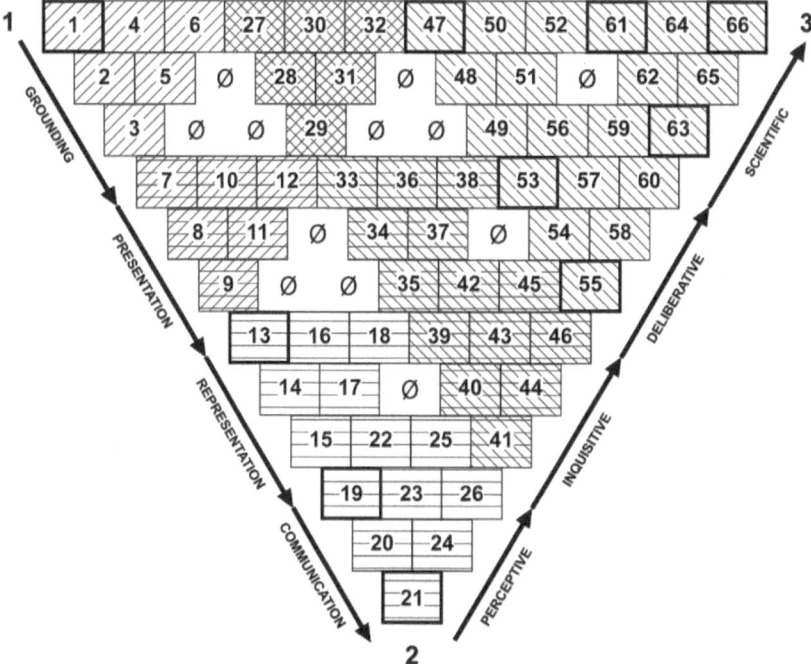

FIGURE 13.6 The remaining sixty-six classes of signs can be arranged in a triangular figure that expands the ten genuine classes of signs presented by Peirce in 1903. The eleven classes eliminated by the razor of causality generate empty spaces.

previous periods, which become invisible to it. Finally, the communication period involves all the others. The final result is that, in a communication process, the transmission of the form of the dynamic object to the final interpretant through the sign occurs in a way that can appear mechanical and linear, but it is truly nesting virtually infinite sequences. As an example of this kind of process, Peirce used to point to the famous paradox of the race between Aristotle and the turtle.

The same occurs with the phases. The perception is nested in the conscious inquiring of the reality. And the deliberation or participatory action nests the two earlier ones. Finally, the scientific activity of knowledge involves all the previous phases. A researcher is only a whole being whenever that person is able to perceive, inquire, deliberate, and communicate the results of his or her research to a community of scientists interested in the truth.

Example: the birth of Aphrodite

It is difficult to capture the complexity of the processes represented by the Solenoid of Semeiosis. To facilitate their understanding, we offer an illustrative example (Figure 13.6) which, although it does not portray all the logically possible relationships described by the solenoid, at least provides the reader with a heuristic path for beginning a study of semeiotic processes. The figure shows a bottom-up process of differentiation, information, and communication that has nothing to do with the top-down mechanical action of an engineered process.

Let us assume that a Greek sculptor[7] has received an order to create a sculpture of the goddess Aphrodite. Figure 13.7 represents the total actions of the sculptor from the moment he begins to produce non-conscious hypotheses (perceptive abductions) about what will become the sculpture, up to the moment when the finished sculpture is disclosed to the audience.

The illustration shows a four-story building hierarchically organized to represent the four periods of the solenoid of semeiosis. In fact, the first story of the figure does not correspond to a floor, but is a kind of basement. It is in the dark, in order to indicate that the processes occurring there fall short of consciousness. It is also important

FIGURE 13.7 When envolving argumentative symbols, semeiosis is a self-organizing bottom-up process guided by rational purposes.

to understand that the actions represented on the four floors can all happen at the same time. For the example to work best, we shall imagine the possibility that the figures of the sculptor work simultaneously, overlapping two or more floors.

The rule for such overlapping is very simple: if the sculptor is active in the grounding period (underground, in our figure), then there will eventually be activity on the immediately upper floor – and the same is valid for higher floors. Thus, if there is no activity during the grounding period, all the remaining floors will be equally inactive. If there is activity in the communication period (on top of the building), then there will necessarily be activity on all the floors below.

Fermenting the dough

The phase of perception occurs in the dark underground (grounding period), which is the production of habits from the qualities of feelings. Here occurs what Peirce describes as collateral experience: the perceptive contact with the object to be represented, generating information. The sculptor must have in the unconscious whirlwind on the back of his mind the myth of Aphrodite, his ideal of feminine beauty, his experiences lived with women, etc.

The dough is shown to the eyes

During the presentation period, we see the emergence of the second floor (consciousness), the constitutive substrate of the representamen (the sign in itself), its own materiality. Once revealed, the sign exposes the type of connection it has with its dynamic object and the effects it is able to produce in reality, regardless of the representation it effectively comes to accomplish. Here occurs the inquiring phase, when the sculptor "get his hands dirty" in order to discover which qualities intrinsic to the substrate can be valued and highlighted for the representamen to optimize its ability to work as a sign.

Giving form to the dough

During the representation period, we have the sculptor acting directly on the substrate of the representamen, aiming to transform it into a representation of Aphrodite. In this period, the deliberation phase occurs, when the sculptor makes choices about how to work the dough, having an idea of the model (the form synthesized in the grounding period and in the perception phase). With patient and dedicated work, the sculptor transfers the form of the object to be represented to the dough/representamen, allowing the potential capacity of representation to be actualized.

Sharing the work with the audience

In the communication period, the forms of the work representing Aphrodite are exposed to public judgment. The exhibition of the work produces dynamic emotional effects (lively discussions on the relevance of a representation of a life-size naked female body, for instance) and, at last, the possible establishment of a generalized opinion on the cultural value of the work.

Conclusion

A semiotic theory of Self-Organization must be sufficiently general to cope with all the processes where information is present. However, a theory of this kind must be a theory of reality, since the real is precisely what is developed by the acquisition of habits in an "in-formational" process (De Tienne, 2005) or, putting it another way, by the internalization and development of the forms. The law of the mind is the very law of the Self-Organization processes.

The conclusion is that laws are not provided a priori to our universe in a mysterious way or by a divine entity, but instead, they arise and gain strength in a semeiotic process where natural arguments (such as the laws of nature) govern the emergence of organization in a way that is similar to biological evolution. This enables us to understand the famous Peircean saying that, "the universe is permeated by signs, if not constituted only by signs" (CP 5.448 n1). These prolegomena to a semeiotic theory of Self-Organization open an important page in the studies of Self-Organization, one that remains to be filled in.

Notes

1 Peirce used several spellings for his doctrine of signs, such as semiotic, semeiotics, and semeotic. A similar situation occurs with the spelling of semeiosis. Here, we adopt the spellings that according to Max Fisch (1986) would be the best for representing the concept that Peirce sought to develop from his reading of Locke. For a rebuttal of Fisch and an alternative view, please see Deely (2008).

2 On natural classes establishing the basis for classifications, see EP 2, pp. 115–132 (Peirce, 1998 will be henceforth referred to as EP 2).

3 This retroductive method is discussed in the conferences published in *Reasoning and the Logic of Things*, pp. 140–141.

4 For the definition of phaneron, please see CP 1.284 and EP 2, p. 362 (both from 1905). For a detailed treatment, see De Tienne (1993).

5 In a letter to Keyser (Peirce, MS 233, 1908), Peirce explains that semeiosis depends on a reality where two parts differentiate without becoming absolutely different: the Psi part is constituted of the habit of breaking habits, while the Phi part is constituted as a habit of forming habits. The Psi part is in charge of the dynamism and flowing of time, while the Phi part produces habits that trend to stiffen, as it is the case of the laws of nature.

6 There is an important observation to be made here: note that the S–OD–ID trichotomy participates both in the objectivation axis and in the interpretation. Therefore, it is enough that the S–OD–ID is a secondness to assure secondnesses to both axes. Likewise, the S–OD–IF trichotomy participates in the three axes. Therefore, the occurrence of the secondness in the trichotomy automatically complies with the rule of the razor of causality.

7 The inspiration for this example was the creation of the first sculpture of a feminine frontal nude by the Greek sculptor Praxiteles (395–330 B.C.), a contemporary of Aristotle. Praxiteles used as model the courtesan Phryne, considered the most

beautiful woman in Greece at that time. The piece was refused by the rulers of the city of Kos, as it was considered obscene, and was exhibited in Knidos where it became famous.

References

De Tienne, A. (1993). Peirce's definition of the phaneron. In: Moore, E. C. (Ed.), *Charles Sanders Peirce and the Philosophy of Science: Papers from the Harvard Sesquicentennial Congress*. Tuscaloosa: The University of Alabama Press, p. 279–288.

——— (2005). Information in formation: A Peircean approach. *Cognitio*, 6(2), pp. 149–165.

Deely, J. (2008). Clearing the mists of a terminological mythology concerning peirce. Available at: www.cspeirce.com/menu/library/aboutcsp/deely/clearing.pdf

Fisch, M. H. (1986). *Peirce, Semeiotic, and Pragmatism: Essays by Max Fisch*. Edited by Kenneth Laine Ketner, and Christian Kloesel. Bloomington: Indiana University Press.

Maturana, H., and Varela, F. (1980). *Autopoiesis and Cognition*. Dordrecht: Reidel Publishing Co.

Peirce, C. S. (1958). *Collected Papers of Charles Sanders Peirce*. Edited by Charles Hartshorne, Paul Weiss, and Arthur W. Burks. Cambridge, MA: Harvard University Press, 1931–35 e 1958, 8v. (CP)

——— (1998). *The Essential Peirce: Selected Philosophical Writings*. V. 2. Edited by the Peirce Edition Project. Bloomington: Indiana University Press (EP2).

Romanini, V. (2006). *Semiótica Minuta – Especulações sobre a Gramática dos Signos e da Comunicação a partir da Obra de C.S. Peirce*. PhD thesis, ECA/USP. Available at: www.minutesemeiotic.org

Scerri, E. R. (1998). *The Evolution of the Periodic System*. Scientific American, September Issue, pp. 56–61.

14

HABIT FORMATION AND SELF-ORGANIZATION

A Peircean approach

Ivo A. Ibri

Introduction

This chapter aims to reflect on the conceptual connection between habit formation and Self-Organization, using Peirce's philosophy as its conceptual ground and discussing specifically his genetic ontology. By genetic, I mean a set of hypotheses intended to explain the origin of all phenomena, and by ontology, I mean the set of hypotheses about the reality of such phenomena. A Peircean approach to this subject primarily calls on an awareness of Peirce's system of interconnected doctrines that refuses every anthropocentric core for conceiving philosophy. Instead of such a core, Peirce proposes a monistic approach that not only ruptures with all genetic dualism but also semiotically extends the properties of the human mind to all-natural phenomena. With this basic foundation, it will be feasible to think of the phenomena of habit formation in every cosmic being, and of Self-Organization as the building of mediations to successfully guide actions towards any deliberated ends, connecting both under the monistic hypothesis of a genetic tendency of mind.

Peirce considered three kinds of induction, which he called *crude, quantitative*, and *qualitative*, and whose description covers cases of experiences strictly focused on scientific investigation. However, it seems that many phenomenological cases aside from scientific inquiry should be also considered, namely, those that constitute the ground of habit formation, taking the notion of *habit* as a rule of conduct formed by generalization or, in general terms, by *induction*. Following this approach, I propose to reflect on the set of semiotic interpretants proposed by Peirce, with the aim of exploring their habitual aspect. These interpretants are not simply theoretical *hermeneutic* signs but, from a pragmatic point of view, should be taken as directly connected with the way some interpreting mind is able, or tending, to act. In such an approach, scientific inquiry is mainly connected with logical interpretants, whose objective is to provide habits of interpreting specific phenomena under some correlated theoretical framework taken as being true. There are, however, some five other types of interpretants we may consider as habits of action. Emotional interpretants, for example, are connected with habits of feeling, similarly formed by induction, that may predominate in certain experiential life situations far from scientific experimental fields. These primarily feed the other three kinds of beliefs that constitute the range of Peirce's method of the fixation of belief, namely, authority, a priori, and tenacity. With such line of analysis, I intend to show that induction is a more generalized operation of mind, far outside the range of the kinds of induction proposed by Peirce. In doing so, I will perhaps be adhering more faithfully to Peirce's very ample philosophical concepts than perhaps he himself intended to be.

On logical and emotional interpretants

My focus in this chapter came from a reflection on Peirce's classic *Fixation of Belief* (CP 5.358–387; EP 1.109–123; W 3.242–257).[1] Given that beliefs are habits of action resulting from inductive generalization, it is appropriate to ask which types of induction apply to each one of the four classes of fixation of belief, namely: scientific, authority, a priori, and tenacity. One can see that the scientific fixation of belief flows from the process that Peirce calls an *inquiry*, that is, in consequence of the triad of hypothesis, deduction, and induction. In this case, the validation of belief happens through the verisimilitude of the hypothesis, from whence experientially observable consequences are validated through induction. This type of induction, which legitimates a scientific argument and establishes a habit of interpretation of the phenomena that are pertinent to it, Peirce calls statistical induction:

> Statistical Induction... assigns a definite value to a quantity. It draws a sample of a class, finds a numerical expression for a predesignate character of that sample and extends this evaluation, under proper qualification, to the entire class, by the aid of the doctrine of chances.
>
> *(CP 7.120–1903)*

Statistical induction has well-defined mathematical models, based on the theory of probabilities, that allow one to estimate average parameters as well as the general distributive dispersion of possible results (e.g., Gaussian functions). Thus, propositions of a probabilistic nature generally take into account, based on the estimation of basic parameters such as mean and standard deviation, the evaluation of probable error associated with any other possible evaluations of parameters and results. It is worthwhile to note that in Peircean epistemology, probabilistic models applicable to phenomenological objects are signs that incorporate degrees of uncertainty stemming from the admission of the reality of chance. This principle, combined with that of law, produces events that define the ontological indeterminism typical of Peirce's realist philosophy (for more details see Ibri, 2017a, Chapter 3).

Scientific beliefs are nourished through a permanent and necessary *semiotic dialogue* between theory and experience, in the form of constant verification of *adherence* of the theoretical predictions of results to the actual results obtained in a determined class of experimental observations (see also Ibri, 2010b, 2012, 2015). New hypotheses will come forth every time that such adherence is broken, not contingently, but significantly (that is, in a temporally redundant manner), exposing the inadequacy of the theoretical model in the face of new data or new experimental amplitude. Scientific beliefs cannot abstain from this adherence of theoretical predictions to experimental results. This very adherence anchors the constitution of what Peirce calls the *community of inquiry*, since it is this community that experiences the intrinsic otherness of the real facts that are being investigated, beyond the idiosyncrasy of the possible particular interests of the researchers involved. Signs and their general meanings circulate within this community in such a way that leads to the collective recognition of the validity of the theories in question. A semiotic network that sustains the community will always be, in ultimate analysis, weighted by facts observed by the community itself.

Peirce also defines a second class of induction, which he calls *qualitative*. In order for this concept to be better understood, we will first explain his idea of *crude induction*. Crude induction always produces universal propositions, totally inclusive or exclusive, based on the presumption that situations and facts that happened in the past will be repeated without exception in the future. Let us consider what Peirce says in this regard:

> The first and weakest kind of inductive reasoning is that which goes on the presumption that future experience as to the matter in hand will not be utterly at variance with all past experience.
>
> *(CP 2.756–1905)*

> I call this Crude Induction. It is the only Induction that concludes a logically Universal Proposition.
>
> *(CP 6.473–1908)*

By qualitative induction, Peirce conceives a logical form situated somewhere between the statistics used in scientific inquiry and crude induction. The validation of a hypothesis by qualitative induction is not given by the statistical calculation of cases, whose method defines a probability and parameters associated with degrees of certainty of the propositions. Validation is configured by the examination of experience under certain qualities of similar cases already available in the repertoire of the inquirer. It seems that there is here a possible association with perceptive judgments based on habits, as occurs in medical diagnostics, for example. Let us consider a passage by Peirce that discusses this type of induction:

The remaining kind of induction, which I shall call Qualitative Induction, is of more general utility than either of the others, while it is intermediate between them, alike in respect to security and to the scientific value of its conclusions. In both these respects, it is well separated from each of the other kinds. It consists of those inductions which are neither founded upon experience in one mass, as Crude Induction is, nor upon a collection of numerable instances of equal evidential values, but upon a stream of experience in which the relative evidential values of different parts of it have to be estimated according to our sense of the impressions they make upon us.

(CP 2.759–1905)

The fallible character of scientific propositions seems to be partially maintained in qualitative induction. This is the very property that it shares with statistical induction. Nonetheless, as it is based on certain impressions that experience mobilizes in the inquirer, it shares this aspect with crude induction. The absence of any example in the Peircean text makes his idea about this type of induction somewhat vague, inducing those who study Peirce to make conjectures about how this would occur in practice. That is why it seems plausible to suppose, as noted earlier, that it can be associated with habitual perceptive judgments.

Once again, I turn my attention to crude induction. This concept seems to be of special interest if associated with three of the four types of belief enumerated by Peirce in his classic *Fixation of Belief* (CP 5.358–387; EP 1.109–123; W 3.242–257). I mention only three of the four types in order to clearly separate scientific belief from the practice of crude induction that, according to Peirce, is always conducive to universal propositions. Scientific propositions, in the light of the ontological indeterminism and epistemological fallibilism of Peircean philosophy, should incorporate a margin of error and of deviation in relation to parameters of higher frequency.

Thus, my main hypothesis here is that the remaining kinds of belief, namely, authority, a priori, and tenacity, are in some way associated with crude induction and consequently produce universal propositions.

In light of the vocabulary of semiotics (for a detailed account of Peirce's semiotics, see SS) it is important for the scope of this chapter to address logical and emotional interpretants inasmuch as we are able to consider them as more sharply distinct in nature, as they relate to time, and in their role as mediations. It may be said of both these interpretants that they are related to energetic interpretants, since they can incur in some form of action (notwithstanding with different ends). Nonetheless, only logical interpretants can be associated with dynamic and final ends, because they are often related to temporal continuity and may be rationally predictive and teleological. Emotional interpretants, when merely qualisigns, are continuities without any logical form and, in this way, constitute genuine possibilities under pure firstness. (Along with secondness and thirdness, firstness is a Peircean categories of experience – about these three more will be said later.) Associated with some factual event, and therefore, under an experience of secondness, emotional interpretants can predominate; in this case, they may cloud that essential categorial otherness able to create a state of doubt, which in Peirce's philosophy, as we know, is potentially the starting point of a new inquiry. It is also possible to reason that emotional interpretants may, in the face of some state of doubt and to the degree in which they predominate in judgments, bring one to universal conclusions without a semiotic dialogue with the temporal conduct of the object. Both cases will be considered later.

The phenomenology of qualisigns and sinsigns

Phenomenologically, we can have pure experiences under the two categories that do not involve objective time, namely, firstness and secondness (for more details, see Ibri, 2017a, Chapter 1). The pure experience of firstness is consummated in the form of free contemplation of the qualities of an object, be it natural or not (see Ibri, 2009, 2010a). A beautiful and serene landscape or a beautiful work of art, or even an engaging piece of music, can grant an experience of unity of conscience defined by a quality of feeling endowed with continuity. In the conception of this experience, there is no notion of finitude or of self-consciousness because of a separation – typical of secondness, between the mind that feels and the object of feeling. Schopenhauer (see Ibri, 2009, 2010a) calls this an esthetic experience, characterized as a *losing of oneself in the contemplated object* to the point of forgetting oneself. Unity or totality, or even continuity, is what this experience contains as a predicate.

Peirce, in a passage in which he reflects on the esthetic good, affirms:

> In the light of the doctrine of categories I and whatever does this is, in so far, esthetically good, no matter what the particular quality of the total may be. If that quality be such as to nauseate us, to scare us, or otherwise to disturb us to the point of throwing us out of the mood of esthetic enjoyment, out of the mood of simply contemplating the embodiment of the quality – just, for example, as the Alps affected the people of old times, when the state of civilization was such that an impression of great power was inseparably associated with lively apprehension and terror – then the object remains none the less esthetically good, although people in our condition are incapacitated from a *calm esthetic contemplation of it.*
>
> *(CP 5.132; italics added)*

This totality or unity of the quality of feelings, which is indeed a continuum of qualities, may be considered pure qualisigns. Signs of this nature do not contain a logical form, and as such, do not possess the teleological characteristics that would make them linked to objective time, which we could call here *Cronos* (Ibri, 2016). This link with time, which makes it such that signs be thinkable as possibly linked to real forms, is typical of signs of thirdness, or legisigns.

Maintaining unity as their essential characteristic, their *whole* being integrated from qualities in a continuum, qualisigns are not associated with any values that could denote them as good or bad: "I am seriously inclined to doubt there being any distinction of pure esthetic betterness and worseness. My notion would be that there are innumerable varieties of esthetic quality, but no purely esthetic grade of excellence" (CP 5.133).

It is worthwhile to mention that, associated with what may be considered the most complex theme within Peirce's conceptualization of normative sciences, *esthetic goodness* cannot, therefore, be grounded only by the nature of qualisigns; this brings one to reason that, given the intrinsic quality of feeling of esthetic goodness, it should contain some logical form that facilitates its being thinkable as an end in and of itself.

Nevertheless, this is not my focus here. We are merely passing through the normative sciences in order to conceive qualisigns as continua of qualities, and thus as continua of possibilities that are phenomenologically experienceable and without logical form.

As such, qualisigns constitute, moreover, one of the main concepts of Peirce's cosmology within his realist ontology. Let us remember that, from this viewpoint, the logical forms of the universe, its habits, its laws, its real third-ness, all have their origin in continua of qualities, which constitute the interior side of the first category (for details about Peirce's cosmology, see Ibri, 2017a, Chapter 5). This is the richest heuristic hypothesis, in my view, of Peirce's philosophy – one which, for example, has implications for the synthetic char-acter of abduction, whose origin is the unity of the perceptive judgment, in which it would be a mixture of qualisign and legisign. The logical form and the unity of feeling are associated in the processes of discovery and invention.

On the other hand, a combination of qualisigns and legisigns also forms the masterpieces of art. In light of this Peircean heuristic, the perception of a new form in science and art, whether as discovery or as a creation still in the abductive stage, happens through a feeling of unity that announces a new form. This phe-nomenology of discovery, I believe, seems to have inspired the cosmogenesis of Peirce's categories, from the generality of possibilities to the generality of logical forms. The laws of nature spring from a tendency towards generalization, the principle *law of the mind*:

> But if the laws of nature are the result of evolution, this evolution must proceed according to some principle; and this principle will itself be of the nature of a law. But it must be such a law that it can evolve or develop itself. Evidently it must be a tendency toward generalization—a generalizing ten-dency... Now the generalizing tendency is the great law of mind, the law of association, the law of habit-taking ...Hence, I was led to the hypothesis that the laws of universe have been formed under a universal tendency of all things toward generalization and habit-taking.
>
> *(CP 7.515)*

Laws are habits of conduct composed of logical forms that are continuous in time. In this abduction of Peirce, they are, in fact, the evolutionary results of thirdness from firstness, passing through the theatre of reactions of secondness. In this vector of evolution, according to Peirce, there is already a direction towards the logical forms of thirdness, where the first category as a continuum of possibilities cohabits with the second category as a mode of being of those existents represented in the relations of law.

Peirce's cosmology, whose conjectural configuration is quite harmonious with the whole of his philosophy's system of doctrines, shows that the logical process of scientific induction happens in reality. Moreover, it is none other than this argument that sustains Peirce's objective idealism – *mind* constitutes the very nature of the whole universe, for the tendency to generalization, its principal law, is the driving axis of evolution (for details about Peirce's objective idealism, see Ibri, 2014, 2017a, Chapter 4).

Legisigns contain qualisigns in their agency as sinsigns. Sinsigns are a kind of sign whose nature and meaning have an existential connection with its object, as, for example, some dark clouds announcing the possibility of rain. They make up the interactivity of the three categories. Qualities and logical forms operate in conjunction with the third category. The legisigns constitute *continua* of both.

One could say that the logical interpretants are always associated with emotional interpretants, and that this synthesis of both constitutes a thinkable unity of time in the form of the schema of Kant, which is valid, one could affirm, in the form of diagrams as Peirce himself conceived them. Perceptive judgments are good examples of this synthesis of time that abductive insights allow (Ibri, 2006). The conjunction of logical and emotional interpretants in judicative diagrams operate in the development of what could be called *sensitivity*, in contrast to pure *emotionality*, which is treated in what follows.

Conjectural judgments should not result in universal sentences. With regard to this feature, there is an ontological abyss between Popperian falsificationism and Peircean fallibilism. While the former searches for exceptions capable of invalidating universal sentences, the latter considers deviations that are consequences of probabilistic laws. Such laws are expressed by the distribution of probabilities that harbor a multitude of possible results with varying degrees of uncertainty. The very nature of conjectural judgments is investigative and, thus, is genetically committed to its own observational character with respect to the otherness of their objects. They establish a semiotic dialogue with experience in order to constitute genuine and reasonably verisimilar representations. Let us consider these consequences for the kinds of belief that are not scientific.

Emotional interpretants and crude induction

I think that it is possible to understand, based on Peirce's realist philosophy, that mediations (Ibri, 2012) are logical forms that represent the general conduct of their objects, and whose function is to make reasonable the cohabitation of individuals that are involved in some sort of relation. By cohabitation, we mean the forms of relation that combine ends endowed with continuity, eliminating binary relationships that tend to remain as such. This is, along general lines, what Peirce's third category promotes; thirdness, which contains secondness, extends the binary relationships into ternary and systemic ones. Mediations should have a logical form that is capable of semiotic dialogue, committed to a necessary epistemological fallibilism, in order to incorporate changes in conduct, the rupture of habits, and the phenomenological diversity that disqualifies universal final propositions of a dogmatic nature. Carrying out these requisites imposes a permanent distinction between the immediate object and the dynamic object, in the recognition of the otherness that should serve as the limit for the form of possible *sayings* about itself (Ibri, 2010b).

When a language destined for the construction of mediations has no anchor in real objects, it always runs the risk of getting lost in judicative arbitrariness, where fiction and reality are no longer distinguished. *Power,* in this case, can only be a term worshiped in its *noun* sense, overlapping its possible meaning as a *verb*, which is provided by the circulation of signs that aim at the joint and potentially communal decipherment of the future course of experience. Well beyond the mere practice of the sciences, scientific belief means a disposition to learn (CP-1.43; 5.582) in a pragmatic sense. This disposition supposes that one maintains the ability to influence conduct in a distinct way, whenever experience demands it.

The three remaining kinds of belief (authority, a priori, and tenacity), as Peirce expounds them, have in common to a greater or lesser degree what I have called in a recent paper *the twilight of reality* (Ibri, 2017b), in other words, a gradual concealment of the dynamic objects that would be able to limit judgments about them. Peirce mentions (CP 5.379–382; EP 1.117–119; W 3.250–252), as an example of belief by authority, the action of the state in some tyrannical manner, imposing its values and ideology on its citizens. This is a typical case in which *truth,* in its sense of a correspondence with real things, is arbitrarily imposed through discourse in its role of exercising *power as a noun*. Language, in this instance, serves secondness only through the exercise of force, being reduced to a mere degenerative thirdness.

A priori belief, as a disposition to believe in that which one is inclined to, is also sealed off from dialogue with experience, making it so that language is reduced to a mere rhetoric that seeks to justify the validity of the belief. When Kant limited scientific discourses to those situated in the realm of possible experience, he was doing no less than maintaining the necessary dialogue with the otherness of phenomena. He clearly distinguished cognitive theories from dogmatic ones – the latter refers to some type of exterior world to which all phenomenological access is sealed. One could say, therefore, that the gradual concealment of reality in kinds of belief makes the dogmatism that typifies them grow in similar proportion.

A priori beliefs encompass a considerable variety of types, intersecting with a belief by authority. Another illustrative example can be drawn from religious beliefs, which, notwithstanding their non-dialogue with experience, institute diverse communities based on distinct interpretations of so-called sacred texts. They almost all share a moral meaning of existence, consoling spirits in the face of the torment of finitude and promising a transcendent correction of earthly injustices.

The most pointed concealment of reality, we may say, occurs in the beliefs by tenacity.

Tenacious is the individual – or even the collective – mind that maintains its opinion unshaken by completely sealing off access to experiences that, by force of their otherness, might be able to provoke ruptures in habits of conduct. Here, once again, language has the inglorious task of justifying what possibly has no mirroring in the world, since through tenacity the immediate objects are distanced absolutely from the dynamic objects.

Allow me here to adopt the term *sensitivity* as a faculty that would operate through mediative perception – notwithstanding its quasi-immediate character – and that flows from the harmonious junction of emotional and logical interpretants. I distinguish this faculty from what could be called *emotionality*, which would operate judicatively by simple apparent factual similarity, separated from the logical forms that have temporal extension, namely, those that preserve their predictive function of future phenomena.

It seems noteworthy, in relation to the three kinds of belief with their respective degrees of dogmatism, to identify a distancing between the logical and emotional interpretants, breaking up their mutual work. In this work, the logical interpretants confer rational direction for judgment, while the emotional interpretants give the proper unity of the quality of feeling that accompanies all good logical form. I take it that the continued exercise of this mutual work grooms what I have here suggested to be called *sensitivity*.

Emotionality, on the other hand, highlights above all the emotional interpretants, super-elevating them above the logical interpretants and confining the latter to rhetorical exercises in language. Separated from their logical counterparts, because of their immediacy and unity, emotional interpretants identify in the facts only the similarities of qualities of feeling, and they tend to construe judgments by mere crude induction, incidents, as Peirce called them, in universal propositions.

This would be an explanatory hypothesis, it seems, for the diverse forms of degenerative thirdness, since judgments solidify in forms of emotional induction that operate by mere analogy with qualities of feeling associated with the secondness of facts. The immediacy of judgments by mere emotional interpretation is not anchored in the temporal conduct of objects, but only in that which their immediate appearance evokes as similar to the immediate qualities of anterior cases.

Evidently, judgments that come from emotional interpretants with no connection to logical interpretants cannot be conjectural; neither can they incorporate any fallible character. Closed within their universality, they do not arouse interest in the future course of the object in the mind that conceives them, something which would be feasible if some logical diagram endowed possible inquiry with signic instruments capable of representing the conduct of these objects.

Self-Organization and habits of interpretation

By Self-Organization, we mean a system of relationships that generate actions teleologically directed towards the interest of the elements that participate in that system, consummating a tendency towards order based on an active interaction of the signs that circulate among those same elements. In light of this concept, we can say that habits of action are self-organized systems arising from the generalization of successful experiences in relation to the desired ends, formed by a natural tendency towards the construction of mediations in relation to any environment with which the system must cohabit.

In essence, it seems legitimate to say that success in achieving the desired ends establishes habits of action originating from induction of a logical character, that is, of statistical content as conceived in the light of Peirce's philosophy. This logical character is justified by the predictive efficiency of the habit – indeed, it constitutes the feedback that guarantees its permanence as mediation. The successful character of prediction implies, it seems sufficiently clear, that habit is inserted in a logical network of a temporal nature. This makes it a genuine mediation.

It is true that mediations of non-logical natures, such as emotional interpretants, are not inserted in the flux of objective time, Cronos, since they do not dialogue with the otherness that participates in real continua, which are of the nature of law. Emotional interpretants are also immediate; if they persistently predominate in judgment, they are not capable of overcoming the mere secondness in which they are immersed and do not reach the third category – it is not in their nature to do so. This passage from secondness to thirdness would be a form of the *discrete* particular becoming generalized as a logical *continuum*. The judgments that solidify or become fixed in mere secondness are, according to Peirce, degenerated. Degenerateness, in this sense, is nothing more than the inability to reach the generality of the third category by means of statistical induction, which is the logical fabric with which logical interpretants are associated.

Thus, we may say that self-organized systems necessarily have a logical nature, and that their interpretive power over the environment in which they are immersed implies a joint labor of the logical and emotional interpretants that constitute them. The logical interpretants give rational direction to the ends that the system seeks to attain. The emotional interpretants confer unity on the judgments, capable of making them quasi-immediate modes of perception of the signs of the environment in which the system is immersed. I mention, here, the expression *quasi*-immediate, since the perceptions are mediated by repertorial signs and not mere sense intuitions. There is, therefore, in these systems the association of two continua, those of an emotional nature and those of a logical nature.

It is interesting to reflect on natural self-organizing systems, bringing one to conclude that ecological equilibrium happens, supposedly, through a very broad communicative network of a logical-emotional nature to which belongs the interchanging of vital signs in temporally efficient habitual processes. Would it not be licit, therefore, to consider the faculty of instinct in the animal kingdom, for example, as a quasi-immediate perceptive-judicative capability in which the logical repertory is constituted by efficient habits acquired by the species? In this sense, would not such a faculty be framed by legisigns and qualisigns forming logical-emotional interpretants whose interactive network constitutes the natural self-organizing system?

It is possible to say that natural beings, except for man, could not adopt crude inductions in order to consolidate habits of conduct, since these by nature do not dialogue with environmental otherness. Their inefficiency as mediations would be fatal. Nonetheless, there should be some margin of error in the action of natural beings, inasmuch as this margin constitutes, as is known, only a minimal fraction of the totality of their respective existences.

Under this aspect, it is also interesting to think about how we human beings are quite frequently submitted to a high degree of dispersion in relation to the correct action towards desired ends. Could we not then conjecture that this characteristic is a consequence of excessive noncommitment between emotional and logical interpretants, making the former distance themselves from the latter and thus promoting a predominance of crude inductions with no predictive power?

It is true that human civilization and culture have brought about a myriad of mediations that offer vital protection and the partial neutralization of our erring based on *continua* with no logical direction, non-dialoguing with otherness and, consequently, separation from Chronos. If a natural being acted in this tenacious way, for example, it would succumb to the secondness of its surrounding environment.

It is important here, nonetheless, in the face of the broadness and even the allure of this theme, to at least suppose that self-organized systems are constituted by the harmonic labor of a judicative competence that acts as its mediation with regards to the environment, formed by the equilibrium of emotional and logical interpretants that constitute efficiently predictive habits of conduct.

Conclusion

In conclusion, I intend to propose a distinction between *sensitive* and *emotional* minds. The first have epistemological abilities that are essential for inquiry, since they provide perceptive judgments that give origin to hypotheses. They are, therefore, immersed in objective time and build mediations through semiotic dialogue with factuality. Emotional minds, on the other hand, extract from the contingency of facts those qualities of feeling that seem to them analogous, tending to adopt conduct originated by crude induction. The universal propositions that flow from this type of induction come from the very nature of the continua of qualities, which, once made discrete by factual secondness, maintain their original tendency towards totalization. In this chapter, we have also considered the distinction between emotionality and sensitivity in the realm of self-organized systems, both human and natural, in the face of the legitimate realist extension of the concept of mind to both domains.

Sensitivity, as a faculty of the mind, is associated with continuous growth, to the extent that it is immersed in a semiotic network that engages in dialogue with the otherness of dynamic objects. The faculty of emotionality is unable to dialogue with otherness because it is not nourished by the logical forms capable of representing otherness with verisimilitude. When emotional interpretants are isolated from logical structures, they incur in crude induction and, as already shown, produce universal propositions. They seem to not be able to distinguish some possible diversity in the unity, taking into account that *unity*, as Peirce affirms, is the proper and essential feature of the quality of feeling that constitutes them.

A myriad of variables of a psychoanalytic nature seem to arise based on this distinctive conjecture between sensitivity and emotionality. What could justify maintaining tenacious beliefs, for example? What could justify adopting power games? What emotional interpretants are connected to the consciousness of finitude and to the risks of a future that could denounce the powerlessness of our mediations?

Does not crude induction become only, then, the generalization of absent cases in the form: that which has never happened may be concluded to never happen in the future (as in the already cited passage in *CP* 2.756, 1905)? Merely emotional induction generalizes affirmatively the cases that contain similar qualities of feeling in the facts, inducing the formation of associated habits, chiefly tenacious beliefs.

Concrete reasonability, the final interpretant of Peircean ethics should, I suppose, rely on those minds endowed with sensitivity, so as to be inserted in a semiotic network of agapic evolution.

The practice of solitude should not be justified by the powerlessness to overcome forms of suffering, but rather be the necessary recollection of sensitivity preparing to introduce its contribution into the world to make that reasonableness feasible.

Note

1 In accord with standard practice, we use the following abbreviations for the various editions of Peirce's works: *CP*, followed by volume and paragraph number, refers to Peirce, 1958; *SS* refers to Peirce, 1977; *EP*, followed by volume and page number, refers to Peirce, 1992; *W*, followed by volume and page number, refers to Peirce, 2010.

References

Ibri, I. A. (2006). The heuristic exclusivity of abduction in Peirce's philosophy. In: Leo, R. F., and Marietti, S. (Eds.), *Semiotics and Philosophy in C.S. Peirce* (pp. 89–111). Cambridge: Cambridge Scholars Press.
———— (2009). Reflections on a poetic ground in Peirce's philosophy. *Transactions of the Charles S. Peirce Society*, 45(3), pp. 273–307.

———— (2010a). Peircean seeds for a philosophy of art. In: Pelkey, J., Haworth, K. A., Hogue, J., and Sbrocchi, L. G. (Eds.), *Semiotics 2010: The Semiotics of Space* (pp. 1–16). New York: Legas Publishers.

———— (2010b). Reflections on practical otherness: Peirce and applied sciences. In: *Ideas in Action – Proceedings of the Applying Peirce Conference* (pp. 74–85). Helsinki: Nordic Studies in Pragmatism.

———— (2012). Choices, dogmatisms and bets: justifying Peirce's realism. *Veritas, PU-CRS*, 57(2), pp. 51–61.

———— (2014). The continuity of life: on Peirce's objective idealism. In: Romanini, V., and Fernándes, E. (Eds.), *Peirce and biosemiotics: a guess at the riddle of life* (pp. 33–49). New York; London: Springer Dordrecht Heidelberg.

———— (2015). The ontology of action in Peirce's philosophy. In: Traykova, E., Cobley, P., Yanakieva, M., Kuncheva, R., and Tashev, A. (Eds.), *The Status of Thought in Honorem Professor Ivan Mladenov* (pp. 76–85). Sofia: Boyan Penev Publishing Centre.

———— (2016). Linking the aesthetic and the normative in Peirce's pragmaticism: a heuristic sketch: Charles S. Peirce society 2016 presidential address. *Transactions of the Charles S. Peirce Society*, 52(4), pp. 598–610.

———— (2017a). *Kósmos Noetós: the metaphysical architecture of Charles S. Peirce*. Philosophical Studies Series, vol. 131, Cham, Switzerland: Springer.

———— (2017b). *The twilight of reality and the melancholic irony of brilliant, unlasting success: reflecting on emotional and logical interpretants in Peirce's modes for fixation of beliefs*. The American Philosophy Forum, 2017 Conference, Emory University – Atlanta, USA.

Peirce, C. S. (1958). *The Collected Papers of Charles Sanders Peirce*. Vols. 1–8. Edited by Hartshorne, C., Weiss, P., and Burks, A. W. Cambridge, MA: Harvard University Press 1931–1958. [We refer to this work in the usual manner: CP indicates Collected Papers; the first number designates the volume, and the second the paragraph.].

———— (1977) *Semiotic and Significs: The Correspondence between Charles S. Peirce and Victoria Lady Welby*. Edited by Charles Hardwick. Bloomington: Indiana University Press. [Referred to as SS]

———— (1992). *The Essential Peirce: Selected Philosophical Writings*. Edited by Nathan Houser, and Christian Kloesel. Bloomington: Indiana University Press, V. 1 [Referred as EP, followed by the number of the volume and number of the page].

———— (2010). *Writings of Charles Sanders Peirce: A Chronological Edition*. Edited by The Peirce Edition Project. Bloomington: Indiana University Press, 1982–2010. 8 V. [Referred as W, followed by the number of the volume and number of the page].

15

ORIGIN OF THE COSMOS AND SELF-ORGANIZATION IN THE WORK OF CHARLES SANDERS PEIRCE

Lauro Frederico Barbosa da Silveira

Introduction

The theme of the origin of the cosmos occupies an important place in the work of Charles Sanders Peirce, and his treatment of this topic places primordial theoretical importance on self-organized processes. Peirce's philosophical proposal helps us to understand the importance of the theme, as well as to understand the frame of reference within which it has developed. In the author's own words, as found in an autobiographical text dating from 1987 (Peirce, 1976, v. 1, pp. 1–14), we find the following:

> Thus, in brief, my philosophy may be described as the attempt of a physicist to make such conjecture as to the constitution of the universe as the methods of science may permit, with the aid of all that has been done by previous philosophers. I shall support my propositions by such arguments as I can. Demonstrative proof is not to be thought of. The demonstrations of the metaphysicians are all moonshine. The best that can be done is to supply a hypothesis, not devoid of all likelihood, in the general line of growth of scientific ideas, and capable of being verified or refuted by future observers.
>
> *(Peirce, 1976, p. 7)*

The universe as cosmos was the central object of Peirce's studies, and he brought to its understanding the whole range of his knowledge. Being an investigation into its ultimate intelligibility, the universe is represented above all metaphysically. Despite this, and also because of it, the author remained constantly attentive to all that the various scientific domains could contribute to his investigation, as well as what they would demand in order to accept its results.

Peirce's study of the subject was strictly hypothetical and purposely fallible. A *form* was being built over time in order to characterize this universe, and in such a fashion that it would be manifested in its reasonableness, although it was never intended to be understood as this, and only this, in its true configuration. The very acceptability of Peirce's account is a challenge for the reader.

The development of his hypothesis is deductive, as is the explanation of the relations involved. The deduction of these relations, however, does not add a greater degree of certitude to what is asserted, but makes explicit its intended reasonableness, making public its principle of construction and, consequently, its consistency.

The reality of the universe was indisputable to Peirce as a physicist and man of science. For Peirce, in the attempt to understand this real object, the deductive process brings forth conclusions that can be submitted to the conditions of experimentation; thus, their explanatory power can be verified over time by the entire scientific community.

Concomitant with the development of the author's hypothesis, which is frequently extremely audacious, Peirce's writings on the subject are punctuated with critical observations about the strategy adopted and the value attributable to the assertions that are made. In these texts, there is also no lack of assessments of the state of knowledge of physical and psychic reality. Peirce points out the need to urgently anticipate frames of logical acceptability for an imminent and radical transformation of the representation of this complex reality that the advance of discoveries in both domains were demanding. Countless Peircean texts seem to converge on the development of this problematic, and we will examine some of them here to the extent that they allow us a first look at such a complex question.

In order to organize the reading of the passages from Peirce reproduced below, this chapter intends to focus on the following items: the refusal to accept the unknowable; the evolutionary and self-organizing character of the universe; and the universal principle of habit acquisition and the law of mind.

The refusal to accept the unknowable

The first aspect of the investigation of the origin of the cosmos and of Self-Organization in Peirce's work seems to us to be a systematic refusal to accept as a matter of law, in whatever sense, a domain of the real that is inaccessible to reason. The Kantian *in itself*, or a uniformity of nature that would sustain an inductive process, as postulated by Mill (Peirce, 1976, v. 1, pp. 92–97), are dismissed by Peirce as corresponding to a suicidal attitude toward reason (Peirce, 1976, v. 1, p. 405); certainly not all of the real will be known, but reason cannot impose *a priori* limits on its investigation.

Only brute fact and pure potentiality do not require that we seek any explanation of them. According to Peirce, the brute fact is necessarily ultimate, and potentiality is at the origin of everything. Regularity and law need an explanation, however, because they break the incommunicability of existence and point in the direction of potentiality. It is they that confer reason on reality, and, therefore, our investigation cannot renounce in principle the ability to penetrate them. If a hypothesis about the real leads to the postulation of the unknowable, that is, if the consequences deduced from it are not observable even in the distant future, it should not be adopted.

Peirce's position leads him, on the one hand, to characterize the universe as an organized whole that is phenomenologically and intrinsically intelligible, and, on the other hand, to characterize all knowledge as eminently fallible. The universe and the spirit who interprets it both share the same nature. Spirit, it may be inferred, is a process of constant and universal interpretation. In the phenomenological domain, where all Peircean investigation is located, to be intelligible is to be interpretable; and it is supposed that all intelligibility that requires interpretation presents itself under the mediation of signs.

This requirement, however, does not subject the universe to a naive anthropocentrism, which the Kantian critique had already denounced as an unfounded pretension to reason acting in dependence on sensible intuitions. Peirce understands man as a spirit to the extent that he belongs to the cosmos, but the texture of the latter is of a spiritual nature. This spiritual nature is manifested in various forms, including forms that are not human, and even including forms that do not belong to the realm of living beings. The following text from 1906 (Peirce, 1976, v. 4, p. 551), for example, mentions crystals: "Thought is not necessarily connected with a brain. It appears in the work of bees, of crystals, and throughout the purely physical world; and one can no more deny that it is really there, than that the colors, the shapes, etc., of objects are really there [...] Not only is thought in the organic world, but it develops there. But as there cannot be a General without Instances embodying it, so there can not be thought without Signs [...] it is not merely a fact of human Psychology, but a necessity of Logic, that every logical evolution of thought should be dialogic".

Intrinsically intelligible, the universe will consequently have an essentially semiotic nature. Thus, in a 1903 text (Peirce, 1976, v. 5, p. 119), Peirce says that:

> The universe is a vast representamen, a great symbol of God's purpose, working out its conclusions in living realities. Now every symbol must have, organically attached to it, its Indices of Reactions and its Icons of Qualities; and such part as these reactions and these qualities play in an argument that, they of course, play in the universe – that Universe being precisely an argument [...] The premises of Nature's own process are all the independent uncaused elements of facts that go to make up the variety of nature which the necessitarian supposes to have been all in existence from the foundation of the world, but which the Tychist [a defender of the theory of the foundational presence of objective chance in the world] supposes are continually receiving new accretions. These premises of nature, however, though they are not the *perceptual facts* that are premises to us,

nevertheless must resemble them in being premises. We can only imagine what they are by comparing them with the premises for us. As premises they must involve Qualities [positive potentialities].

Now as to their function in the economy of the Universe, the Universe as an argument is necessarily a great work of art, a great poem – for every fine argument is a poem and a symphony – just as every true poem is a sound argument. But let us compare it rather with a painting – with an impressionist seashore piece – then every Quality in a Premiss is one of the elementary colored particles of the Painting; they are all meant to go together to make up the intended Quality that belongs to the whole as whole. That total effect is beyond our ken; but we can appreciate in some measure the resultant Quality of parts of the whole – which Qualities result from the combinations of elementary Qualities that belong to the premises.

To understand the universe as cosmos is, therefore, to learn to interpret it as a complex argument, discovering its intrinsic principle of formation, having as an ultimate goal the revealing of it in the perfection of its form, which is like that of a poem. Thus, it seems to the physicist Peter Voetmann Christiansen (1993, pp. 223–245) that a satisfactory solution to the question of the possibility of some continuous quantities presenting exact values even if not measured (a subject that provoked a division of opinion among Niels Bohr, Heisenberg, and Einstein with regard to the ontological status of uncertainties in the field of quantum mechanics) could be found in the Peircean theory of the semiotic and realistic character of the universe, due to the fact that it does not limit law and rationality merely to their symbolic relations. In this regard, Christiansen states:

> Peirce could maintain his realism without the notion of exact numerical values of potential observables because he had developed semiotic to a point where it was able to treat classes of signs existing independent of the human consciousness. The index category of signs seemed to have acted as a 'secret weapon' (it was not mentioned directly in the *Monist* papers) that gave him the strength to withstand the nominalistic temptation.

Christiansen continues:

> In general, a sign is conceived as a genuine triadic relation between (1) a sign vehicle, (2) an object, and (3) an interpretant. However, the interpretant may be latent, and in its absence the sign is of a 'degenerate' type, called an index, expressing a dyadic relation between sign vehicle and object. Thus, a physicist is concerned with translating the indexical signs of nature into the symbolic signs of physics, and this process involves the setting of an interpretant. The quantum mechanical measurement process

fits well to this general description with the measuring apparatus as the embodiment of the interpretant.

(Christiansen, 1993, p. 227)

It, thus, becomes unnecessary to decide in an exclusive manner between the reality of the laws of nature, here presented by their indices, and the representation required by the logic of our theoretical discourse. Both semioses are authentic generalities and present forms that are exclusive of one another, without, however, one not being interpretable within the other; both effectively share the authentically poetic nature of the same cosmos.

The evolutionary and self-organizing character of the universe

Having the nature of thought, the universe is understood by Peirce as a process that is genetically in constant evolution towards an ever larger and more encompassing organization. Although this concern was already present in earlier texts, a text from 1891 presents a rather complete picture of this idea (Peirce, 1958, v. 8, p. 317). Referring to his concern with the ultimate constitution of the universe, Peirce writes:

> I may mention that my chief avocation in the last ten years has been to develop my cosmology. This theory is that the evolution of the world is *hyperbolic*, that is, proceeds from one state of things in the infinite past, to a different state of things in the infinite future. The state of things in the infinite past is chaos, tohu bohu, the nothingness of which consists in the total absence of regularity. The state of things in the infinite future is death, the nothingness of which consists in the complete triumph of law and absence of all spontaneity. Between these, we have on *our* side a state of things in which there is some absolute spontaneity counter to all law, and some degree of conformity to law, which is constantly on the increase owing to the growth of *habit*. The tendency to form habits or tendency to generalize, is something which grows by its own action, by the habit of taking habits itself growing. Its first germs arose from pure chance. There were slight tendencies to obey rules that had been followed, and these tendencies were rules, which were more and more obeyed by their own action. There were also slight tendencies to do otherwise than previously, and these destroyed themselves. To be sure, they would sometimes be strengthened by the opposite tendency, but the stronger they became the more they would tend to destroy themselves.

Peirce (1958, v. 8, p. 318) continues:

> I believe the law of habit to be purely psychical. But then I suppose matter is merely mind deadened by the development of habit. While every

physical process can be reversed without violation of the law of mechanics, the law of habit forbids such reversal. Accordingly, time may have been evolved by the action of habit. At first sight, it seems absurd or mysterious to speak of time being evolved, for evolution presupposes time. But after all, this is no serious objection, and nothing can be simpler. Time consists in a regularity in the relations of interacting feelings. The first chaos consisted in an infinite multitude of unrelated feelings. As there was no continuity about them, it was, as it were, a powder of feelings. It was worse than that, for of particles of powder some are nearer together, others farther apart, while these feelings had no relations, for relations are general. Now you must not ask me what happened first. This would be as absurd as to ask what is the smallest finite number. But springing away from the infinitely distant past to a very very distant past, we find already evolution had been going on for an infinitely long time. But this "time" is only our way of saying that something had been going on. There was no real time so far as there was no regularity, but there is no more falsity in using the language of time than in saying that a quantity is zero.

It is clearly possible in this passage to see that the formation of habit is the only principle required by Peirce for explaining the continuous and growing evolution of the cosmos and the expansion of the domain of law, the original precedence of chance and spontaneity over any coercion and over any regularity, and the essential psychophysical unity of the cosmos:

In this chaos of feelings, bits of similitude had appeared, been swallowed up again. Had reappeared by chance. A slight tendency to generalization had here and there lighted up and been quenched. Had reappeared, had strengthened itself. Like had begun to produce like. Then even pairs of unlike feelings had begun to have similars, and then these had begun to generalize. And thus relations of contiguity, that is connections other than similarities, had sprung up. All this went on in ways I cannot now detail till the feelings were so bound together that a passable approximation to a real time was established. It is not to be supposed that the ideally perfect time has even yet been realized. There are no doubt occasional lacunae and derailments.

The universal principle of habit acquisition and the law of mind

The domain of mechanics, in which prevails the law of energy conservation (or the *vis viva*, as Peirce often refers to it) and in which time is reversible, does not constitute the first and most universal law of the universe. In fact, Peirce admits two separate domains in the universe: one strictly mechanical and the other characteristic of mental phenomena.

The physical phenomena explained by classical mechanics must themselves show a minimal divergence from law. This is true even if, in view of the degree of vastness with which they are habitually observed and of the ends for which they are observed, they behave in a strictly deterministic way (Peirce, 1976, pp. 46–54).

Chance is the first principle, and it runs through the entire universe. Existence, and with it the causal relations of action and reaction, concretizes quality. However, this does not suppress its potentiality and spontaneity. If there were no survival of spontaneity, there would be no place for law and for the genesis of evolutionary processes.

Mechanical laws are not properly the laws of the universe, but abstract operative principles that satisfy the representation of certain phenomena for certain ends, within a certain degree of approximation (Peirce, 1976, v. 1, pp. 348–349). The laws that form the universe are much weaker. On the one hand, they clearly characterize the phenomena of the mind; on the other, they can represent physical phenomena if these phenomena are considered to be processes governed by the same evolutionary law of the acquisition of habits, but already in a terminal state of evolution (Peirce, 1958, v. 8, p. 318).

The irreversibility of some physical–chemical properties of matter, as in case of viscosity, provides this hypothesis with hope for its more precise empirical verification. In the viscosity of the protoplasm of nerve cells, and in the long carbon chains of the substances constitute them, Peirce saw the possibility, still remote given the scientific knowledge of his time, of finding the physiological basis of mental phenomena.

The protoplasmic cell reacts to stimuli, manifesting itself as having a sensory capacity; its ability to transmit impulses from one cell to the other makes it capable of action and reaction and of the emission of responses. Finally, the cell's diverse properties of stimulation, inhibition, and fatigue enable nerve cells, acting in a network, to acquire specific learning habits.

Mind and matter, though distinct, are inserted into the continuous flux. Peirce thus affirms that although matter only acts on matter, and spirit on spirit, the continuum that sustains them allows the infinite to be traversed and allows something representable by a singularity that designates the influence exerted by matter over spirit and vice versa (Peirce, 1958).

The following text contains a succinct formulation of the law of mind, which can be understood as the basic law of the constitution of the cosmos:

> First, then, we find that when we regard ideas from a nominalistic, individualistic, sensualistic way, the simplest facts of mind become utterly meaningless. That one idea should resemble another or influence another, or that one state of mind should so much as be thought of in another, is, from that standpoint, sheer nonsense.
>
> Second, by this and other means we are driven to perceive, what is quite evident of itself, that instantaneous feelings flow together into a continuum of feeling, which has in a modified degree the peculiar vivacity of

feeling and has gained generality. And in reference to such general ideas, or continua of feeling, the difficulties about resemblance and suggestion and reference to the external cease to have any force. Third, these general ideas are not mere words, nor do they consist in this, that certain concrete facts will every time happen under certain descriptions of conditions; but they are just as much, or rather far more, living realities than the feelings themselves out of which they are concreted. And to say that mental phenomena are governed by law does not mean merely that they are describable by a general formula; but that there is a living idea, a conscious continuum of feeling, which pervades them, and to which they are docile. Fourth, this supreme law, which is the celestial and living harmony, does not so much as demand that the special ideas shall surrender their peculiar arbitrariness and caprice entirely; for that would be self-destructive. It only requires that they shall influence and be influenced by one another. Fifth, in what measure this unification acts, seems to be regulated only by special rules; or, at least, we cannot in our present knowledge say how far it goes. But it may be said that, judging by appearances, the amount of arbitrariness in the phenomena of human minds is neither altogether trifling nor very prominent.

(Peirce, 1958, v. 8, pp. 274–275)

The continuum presupposed by the law of the mind sustains the cosmic unity in all its degrees of vastness, and the presence of chance confers the necessary spontaneity for this continuum to take place and not break up into a brutal theater of reactions. With the gentle attraction that the similars maintain among themselves, a primordial bond, prior to any shock, is established; Peirce very wisely calls this bond as an *affection*. The evolutionary process, in the form of continuous growth and diversification, confers upon the cosmos its teleological dimension of perfection. In this process, Peirce recognizes the *entelecheia* proposed by Aristotle.

Thus, writing in 1897, Peirce recognizes a deep affinity between his thought and that of the ancient philosophers, despite the distance that separates them. The American thinker, thus, goes on to say that:

But fallibilism cannot be appreciated in anything like its true significancy until evolution has been considered. This is what the world has been most thinking of for the last forty years – though old enough is the general idea itself. Aristotle's philosophy…is but a metaphysical evolutionism.

(Peirce, 1976, v. 1, p. 173)

The law that embodies the principle of the acquisition of habits evolves in this very process. Laws will give way to other laws, for they are nothing more than the unfolding of reason in the universe, the expression of acquired habits. The genesis of time, the diversity of possible worlds, and various other themes, were worked out by Peirce in the light of these principles, always with a view to providing future scholars with eminently philosophical assistance in the risky task of elaborating hypotheses better suited to the hope of an always better understanding of the universe. If such assistance does not seem to have had any effect at first glance, it is nonetheless in accordance with exigencies that remain relevant up to the present day. To understand the universe is to grow in perfection, and this is the ideal present at the heart of reason.

In 1903, Peirce could finish a lecture at Harvard with the following words:

> The creation of the universe, which did not take place during a certain busy week, in the year 4004 B.C., but is going on today and never will be done, is this very developement of Reason. I do not see how one can have a more satisfying ideal of the admirable than the development of Reason so understood. The one thing whose admirableness is not due to an ulterior reason is Reason itself comprehended in all its fullness, so far as we can comprehend it. Under this conception, the ideal of conduct will be to execute our little function in the operation of the creation by giving a hand toward rendering the world more reasonable whenever, as the slang is, it is "up to us" to do so. In logic, it will be observed that knowledge is reasonableness; and the ideal of reasoning will be to follow such methods as must develope knowledge the most speedily.

(Peirce, 1976, v. 1, p. 615)

References

Christiansen, P. V. (1993). Peirce as participant in the Bohr-Einstein discussion. In: Moore, E. C. (Ed.), *Charles Sanders Peirce and the Philosophy of Science* (pp. 223–245). Papers from the Harvard Sesquicentennial Congress. Tuscaloosa and London: The University of Alabama Press.

Peirce, C. S. ([1934] 1976). *Collected Papers of Charles Sanders Peirce*, Vols. 1–6. Edited by Charles Hartshorne and Paul Weiss. Cambridge, MA: The Belknap Press of Harvard University.

———. (1958). *Collected Papers of Charles Sanders Peirce*, Vols. 7–8. Edited by Arthur Burks. Cambridge, MA: Harvard University Press.

16

DIAGRAMMATIC LOGIC AS A METHOD FOR THE STUDY OF COSMOVISIONS

Enidio Ilario

Diagrams and symbolism

The method referred to as the *ordine geometrico* (geometrical order) has always been a widely used resource in philosophy, and has deep roots in the field of human knowledge. For example, before heliocentrism became the model of the modern cosmovision, the geocentric Aristotelian-Ptolemaic theory prevailed for centuries, maintained by canonic force. Even Copernicus, Galileo, Kepler, Tycho Brahe, and other great precursors of the contemporary view of the cosmos were obligated to pay tribute to the naïve geocentric cosmovision that was imbued with astrology. In the modern era, from the macrocosmic Keplerian solar system to the microcosmic atom of Rutherford, there are many examples of the heuristic richness of modeling in the resolution of problems. In the present study, we will examine the use and understanding of geometrically ordered diagrams by a number of important traditions and thinkers throughout history, focusing on the set of relations in these diagrams that establish a structure that is determinant in the explanation of phenomena. We suggest that they share certain characteristics in common that reflect basic human aspirations and the structure of the mind (Ilario, 2011).

Bachelard (2002) argues that scientific knowledge is always the reform of an illusion, but that intuition and creative imagination feed and renew the creative activity of thought. The same author alludes to the difficulties in developing truly innovative thought, and tells us that attempts at geometrization were exceptional and slow in developing (Bachelard, 2002). In the period between the beginnings of Christianity and the Renaissance, the Middle Ages produced the subtle ingredients which – supplemented by the hermetic-Kabbalistic tradition – secured the basis of modern science, which has as its basis the methodical search for explanatory theories.

SECVNDA FIGVRA.

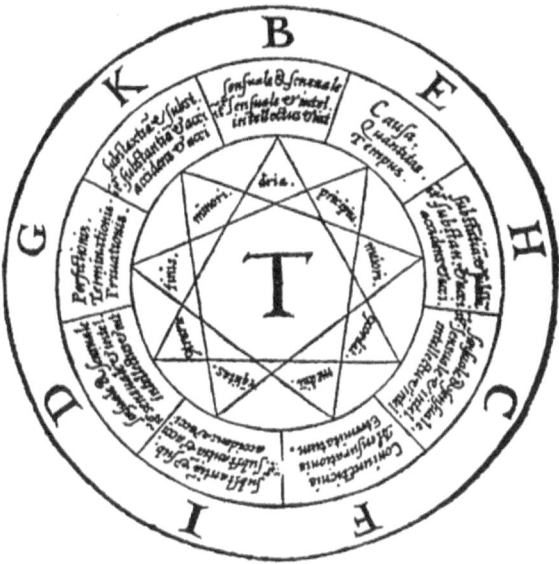

FIGURE 16.1 Diagram of concepts from Llull's *Ars Breve*. Wikimedia Commons.

A thinker who continues to occupy an enigmatic role in the history of knowledge is Ramon Llull (1232–1316), also known as Raymond Lully. The seminal attempt by Lull to construct a diagrammatic model of the movements of the human psyche was expounded in his *Ars Magna* of 1305, and in didactic form in his *Ars Breve* of 1308. Figure 16.1, from the *Ars Breve*, represents concepts by means of letters of the alphabet; the letters are not static but turn about on an axis in such a way as to form meaningful combinations. Lull, a writer, philosopher, theologian, and Catalan missionary, would later become an influence on some of the important figures of the Renaissance.

The complex route that connects these traditions was brilliantly described by Frances A. Yates (1899–1981) in her classic *The Art of Memory* (Yates, 1966), a title which recalls the tradition that began with the pre-Socratics and extends from Plato and Aristotle to the late Middle Ages. As Yates has shown in another work, *Giordano Bruno and the Hermetic Tradition* (1964), a typical phenomenon of syncretism occurred during the Renaissance. This process, in which "the art of memory" had great prominence, combined Greek philosophy, primitive Christianity, and the magic arts that certain Renaissance thinkers mistakenly assumed descended from ancient Egypt and Mesopotamia. According to Yates, the art of memory is a clear case of a marginal theme, and "the serious investigation of this forgotten art may be said to have only just begun" (Yates, 1966, p. 389). Initially suspicious and hostile with regard to the hermeticism underlying the work of Renaissance thinkers such as Gordiano

Bruno and his contemporaries, Yates later recognized that, in fact, a special encounter of traditions had germinated the seed of modern science. With regard to Giordano Bruno, Yates argues that, underneath a certain pathological component, there is in Bruno an intense effort in the search for a method that "can only be described as a scientific element, a presage on the occult plane of the preoccupation with method of the next century" (Yates, 1966).

It is in this spirit that Yates, from the point of view of the history of ideas, establishes a consistent nexus between Lully's desire for a Kabbalistic-ascensional method and Leibniz' constructions of almost four centuries later. Leibniz' efforts to conceive of a universal calculus by means of combinations of signs and meaningful characters is a tribute to the legacy of the hermetic-Kabbalistic tradition. At the same time, his symbols are mathematical and his combinations gave origin to the infinitesimal calculus. As Yates points out (1966), the diagram at the beginning of Leibniz' *Dissertatio de arte combinatoria* of 1666, in which the square of the four elements is associated with the logical square of oppositions, shows that Leibniz understood Llullism as natural logic (Figure 16.2).

Figure 16.3 is a diagram *(Selo)* by Giordano Bruno which portrays the logical square of Apuleius. Within this diagram of the "magic of memory", there is inculcated a logic, perhaps a natural one. Unquestionably, however, one also finds in it what would become contemporary symbolic logic as inaugurated by George Boole in the mid-nineteenth century. Somewhat later, on the threshold

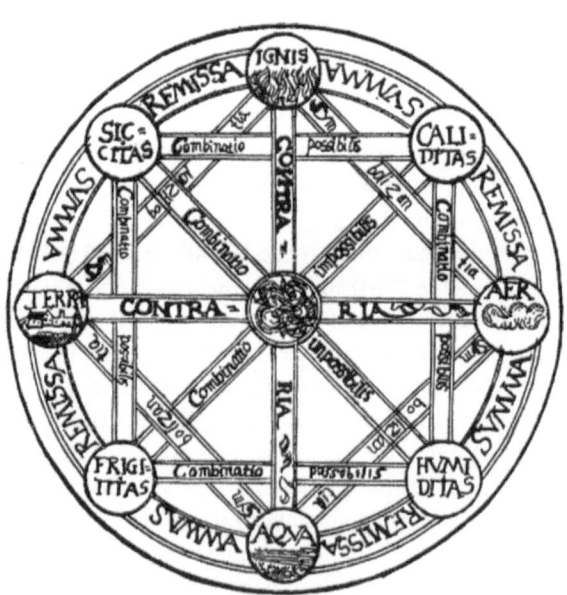

FIGURE 16.2 Frontspiece of Leibniz' *Dissertatio de arte combinatoria.* Wikimedia Commons.

FIGURE 16.3   Logical diagram by Gordiano Bruno. *De umbris idearum* (1582),
Wikimedia Commons.

of the twentieth century, another great logician, Charles Sanders Peirce, would
dedicate the best of his efforts to the development of a diagrammatic logic, on the
basis of which he drew existential graphs.

A passage by Deleuze and Guattari (1994) sums up the spirit that moves
the present study. In their work, the authors refer to archaic symbolism and
correlate the occult meanings in symbols with a plane of immanence that
holds out possibilities for the elucidation of philosophy and of science itself.
Their diagrammatic nature is exactly what such symbolic constructs have in
common: "It is a wisdom or a religion – it does not much matter which. It
is only from this point of view that Chinese hexagrams, Hindu mandalas,
Jewish sephiroth, Islamic 'imaginals', and Christian icons can be considered
together: thinking through figures. Hexagrams are combinations of continuous and discontinuous features deriving from one another according to the
levels of a spiral that figures the set of moments through which the transcendent descends. The mandala is a projection on a surface that establishes a
correspondence between divine, cosmic, political, architectural, and organic
levels as so many values of one and the same transcendence. That is why
the figure has a reference, one that is plurivocal and circular by nature. [...]
And yet disturbing affinities appear on what seems to be a common plane of
immanence. [...] This is because figures are projections on the plane, which
implies something vertical or transcendent. Concepts, on the other hand,
imply only neighborhoods and connections on the horizon" (Deleuze and
Guattari, 1994, pp. 89–92).

We can see that the authors are not referring to any sort of occult (esoteric) art. Deleuze and Guattari (1994) note that it is only with Descartes, and with Kant and Husserl, that it became possible to treat the plane of immanence as a field of consciousness, as immanence is supposed to be immanent to a pure consciousness, to a thinking subject which Kant referred to as *transcendental*. In this context, the subject thinks the concept; this is an act of thought that is always created on the basis of other concepts as a heterogenesis, that is, as an ordering of its components by zones of proximity. The concept, therefore, is an ordinal, a tension, and an intention present in all the features of which it is composed; it possesses a coming into being that concerns its relation with concepts situated on the same plane. In the figures/diagrams discussed here, there are properties analogous to a plane of immanence, which are inter-translatable. In these figures/diagrams, we find the same structure that reflects the diagrammatic nature of the human mind itself.

Regarding the animal symbolicum

In his *Essay on Man*, first published in 1944, the philosopher Ernst Cassirer (1874–1945) coined the expression *animal symbolicum*, arguing that the functional circle of man had not only been quantitatively augmented, but had also undergone a qualitative change in order to be able to adapt to the environment. He recognized the symbol as the key to the nature of man: "he cannot see or know anything except by the interposition of this artificial medium" (Cassirer, 1953, p. 43); "The principle of symbolism, with its universality, validity, and general applicability, is the magic word, the Open Sesame! giving access to the specifically human world, to the world of human culture" (Cassirer, 1953). He continues: "In order to grasp this meaning man is no longer dependent on concrete sense data, upon visual, auditory, tactile, kinesthetic data. He considers relations 'in themselves' – *auto kath' hauto* – as Plato said. Geometry is the classic example of this turning point in man's intellectual life [...] we are studying universal spatial relations for whose expression we have an adequate symbolism" (Cassirer, 1953, pp. 58–59).

Greimas and Cortés (2008), recognizing that the term *symbol* admits multiple definitions characterizing syncretism and ambiguity, do not recommend its use in semiotics. In philosophy and psychology, we would like to make it clear that symbol must not be understood as synonymous with *sign* in general, as the former pertains to a specific class of the latter – that of figurative or iconic signs, that is to say, structural signs. Silveira (2014, p. 134) says that "as a diagram, reasoning will thus eminently be an icon, as a form of intelligible relations. Diagrammatic construction does not lack a place for the symbol and the index". For Greimas, figurative signs are objects that preserve characteristics and similarities with what they represent, in accordance with the culture of those who interpret them (Greimas and Courtés, 2008). In the symbol, there is a formal character that is not necessarily present in the sign, as the sign can be natural (for example, smoke can be a sign of fire); but beyond the formal character of the symbol, we wish to highlight those symbols that contain analogies with what is represented (symbolized).

While scholars of religious or numinous symbolism usually highlight the function of symbols as representative signs of realities not accessible by means of theoretical reason, we wish to show that a specific class of such symbols are themselves perfectly accessible to reason. In this case, we refer to symbols of a geometric nature, with nuances of symmetry and asymmetry, and we wish to point out the diagrammatic character that is common to all of them, as defined by Gardner (1958). With regard to this definition, it is worth asking if it also applies to religious symbolism, that is, that which contains a numinous element.

Cosmogonies and the axis mundi

According to Eliade (1959), man becomes conscious of the sacred because it manifests itself. To indicate the act of such a manifestation, he proposes the term "hierophany". One of the examples given by the author is the phenomena of the *quaternio*, the division of villages and cities of certain peoples according to the four cardinal points. He attributes such a division to the cultural conception of the *axis mundi* (the axis or pillar of the world), which marks the center from which radiate the axes pointing to cardinal points. According to the author, to live in the world it is necessary to ground it, and no world can be born in the "chaos" of the homogeneity and relativity of profane space. Therefore, the discovery or projection of a fixed point, the "Center", equivalent to the creation of the world, is found in human culture. Regarding the "navel of the world", Eliade refers to the Mesopotamian, Judeo-Christian, and Iranian traditions, in which the center is the place where a rupture of levels is effectuated, and where space becomes truly sacred and real (Eliade, 1959).

Eliade further states that is not surprising to find a similar conception in ancient Italy and among the ancient Germans, as we are dealing with an archaic and widespread idea, that is, that the four horizons are projected in the four cardinal directions from a center. Corroborating Eliade's assertion, it is worth citing Gregory of Nyssa's description of the Christian cross from 394 AD: "The cross unites the four cardinal points and thus symbolizes the unity of the cosmos: its north-south vertical axis connects heaven to hell, while the east-west axis covers the earth. It is the 'axis mundi', the 'tree of life'" (Leloup, 2006, p. 73). According to Eliade, "The three cosmic levels – earth, heaven, underworld – have been put in communication [...] the image of a universal pillar, *axis mundi*, which at once connects and supports heaven and earth and whose base is fixed in the world below (the infernal regions). Such a cosmic pillar can be only at the very center of the universe, for the whole of the habitable world extends around it" (Eliade, 1959, pp. 36–37).

Figure 16.4 shows the upper surface of a Shaman drum, whose original design, according to the source, was obtained between 1909 and 1913 during ethnographic expeditions in the south of Siberia and the Altai Mountains. Shaman drums illustrate the pictorial conception of the *axis mundi*, making clear the hierarchic characteristic of the above and the below in a symbolic map of the universe. The space is divided into two important zones: the sky above (the higher world) with the stars, and the human world (the middle world) below the horizontal line. On the left side is the Shaman holding the drum, and above him are mountain goats. On the right side is a horse by a tree, ready to be sacrificed, and above is the same animal after being sacrificed.

Shamanic art is prodigious in the reaffirmation of nature in terms of a hierarchizing visual syntax, and, like Egyptian iconography, is far from being a naïve spontaneous art lacking subtle elaborations and abstractions of cosmic meaning. In this regard, it is worth citing the speech of the celebrated Oglala Lakota (Sioux) shaman Black Elk, who participated in the battle of Little Big Horn in 1876 (a well-known battle in which the Sioux, led by Sitting Bull, inflicted a serious defeat on the United States Army commanded by General Custer): "Grandfather, Great Spirit… You have set the powers of the four quarters [of the earth] to cross each other. The good road and the road of difficulties you have made to cross; and where they cross, the place is holy. Day in and day out, forever, you are the life of things" (Neihardt, 2008). This same conception is also found in an extraordinarily elaborated form in the celebrated Egyptian Zodiac of Dendera.

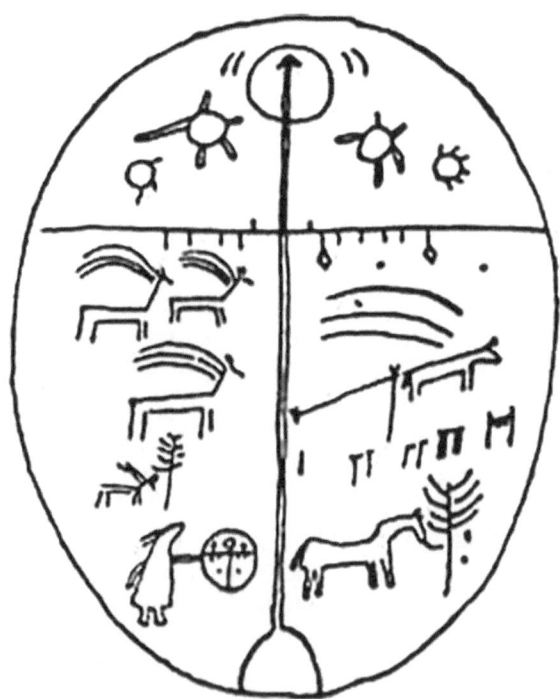

FIGURE 16.4 Shaman drum that illustrates the *axis mundi*. Wikimedia Commons.

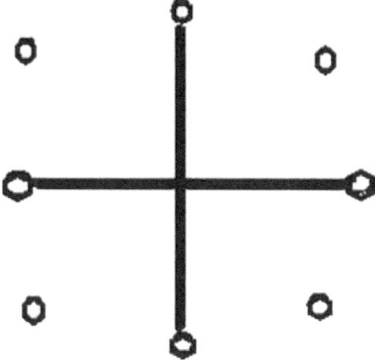

FIGURE 16.5 Ancient rock art diagram; illustration by the author.

Discovered in 1799, during the Napoleonic incursions into Egypt and now on exhibit in the Louvre, this stone bas-relief is a cartography of the heavens based on the constellations of the zodiac. Similarly, a complex celestial iconography can also be seen in the Stone of the Sun, an Aztec calendar of approximately 25 tons, dating from the beginning for the sixteenth century and now on display at the National Museum of Anthropology and History in Mexico City.

Manifestations of the sacred are often structured in oppositions containing the above, the below, the left, and the right. In ancient rock art, we find pictorial manifestations that show the same imagistic dispositions dating back to at least the Mesolithic period and the Bronze and Iron Ages (Coimbra, 2004). An example may be observed in Figure 16.5. Aside from ancient manifestations, pictorial manifestations of the sacred may also be found in the present day in all cultures, as, for example, in the case of contemporary Afro-Brazilian Umbanda rituals (Lima, 1997, pp. 70–82).

Lévi-Strauss: What is the purpose of Kadiweu art?

The anthropologist and ethnologist Lévi-Strauss, in his classic work *Tristes Tropiques* (Levi-Strauss, 1961, 2012), extensively discusses the indigenous Caduveo (or Kadiweu) people of Brazil. According to the author, the tribe is descended from the famous indigenous horsemen of the Pantanal region, the Mbaya-Guaicurus, who were known in the past as great warriors. Kadiweu motifs are complex, geometric, and impressive to the observer. Aside from appearing on the body, the natural location of Kadiweu painting, the motifs appear on hides, mats, and fans. According to Lévi-Strauss, Kadiweu motifs are incomparable: "The face, and sometimes the entire body, was covered with a network of asymmetrical arabesques that alternated with subtle geometrical motifs" (Lévi-Strauss, 1961, p. 164). Chapter 17 of *Tristes Tropiques* (Levi-Strauss, 1961, p. 160 ff.), contains several drawings of Kadiweu designs, and in these, it is not difficult to note the geometric patterns alternating in symmetry and asymmetry, divided in an orthogonal plane, and forming quadrants.

The Belgian ethnographer observed that these Indians "created a graphic art which is quite unlike almost everything that has come down to us from pre-Columbian America, although it does have some similarity to the figures and patterns on our playing cards" (Lévi-Strauss, 2012). In the end, what is their purpose?

> The face-paintings confer upon the individual his dignity as a human being: they help him to cross the frontier from Nature to culture, and from the 'mindless' animal to the civilized Man. Furthermore, they differ in style and composition according to social status, and thus have a social function.
>
> *(Lévi-Strauss, 1961, p. 176)*

In Lévi-Strauss' view, associationist psychology has the merit of having outlined an "elementary logic", but failed to recognize that "it was an original logic, a direct expression of the structure of the mind (and behind the mind, probably of the brain)" (1964, p. 90). The author then goes on to suggest that: "A renovated associationism would have to be based on a system of operations which would not be without similarity to Boolean algebra" (Lévi-Strauss, 1964, pp. 90–91).

Jung and the symbology of the self

Among the greatest names of psychology, and possibly the one who most systematically dedicated himself to the decoding of symbolism, recognizing its importance in the study of the psyche, is Carl Gustav Jung (1875–1961). Jung considers the symbol as the best formulation of an object that is not perfectly identifiable in all of its aspects (Jung, 1968b). The philosophical speculation of educated Europe was attracted to oriental symbols, to the grandiose Indian conceptions of divinity, and to the profound depths of Chinese Taoist philosophy, in the same way that in other times the heart and the spirit of ancient men were captivated by the Christian ideal (Jung, 1968a). Thus for Jung, the symbolic process is one of living in the image and for the image, and shows in its regular development an enantiodromic[1] structure like that of the text of the *I Ching*, presenting a rhythm of negation and affirmation, loss and gain, dark and light (Jung, 1968a). For Jung, the polarities are constitutive of the human psyche, thus his conviction about the substantial existence of Evil and of the corresponding idea of Good. The black and the white, the light and the dark, the good and the bad, are pairs of contraries, one presupposing the other (Jung, 1959). In his classic work, *The Archetypes and the Collective Unconscious* (Jung, 1968a), Jung makes this conception clear on the basis of the conviction that human intelligence decomposes the totality of antinomic judgments: "[…] in all chaos there is a cosmos, in all disorder a secret order, in all caprice a fixed law, for everything that works is grounded on its opposite. It takes man's discriminating understanding, which breaks everything down into antonomical judgments" (Jung, 1968a, p. 66).

According to Jung, relations of the yang-yin type are much closer to the factual truth than the *privatio boni*. In his opinion, the yang-yin conception in no way causes a rupture in monotheism, so that that the yang and the yin, illustrated by the classic *Tai-Ji* diagram shown in Figure 16.6, represent the integrative unity of the *Tao*, which the Jesuits correctly translated as "God" (Jung, 1959).

Jung also proposes "an archetype of wholeness, i.e., the self" (Jung, 1959, p. 223): "The most important of these are geometrical structures containing elements of the circle and quaternity; namely, circular and spherical forms on the one hand, which can be represented either purely geometrically or as objects; and, on the other hand, quadratic figures divided into four or in the form of a cross" (Jung, 1959, pp. 223–224). These are variations on a fundamental theme, the *mandala*, a sacred symbol that in India has the name of *yantra*.

A prime example is the well-known Tibetan Wheel of Life, a typical Tibetan mandala, in which the polarities in the graphic design represent human delights and torments, gods and demons, and the polarity of good (above) and of evil (below). In Tibetan Buddhism, it is more appropriate to speak of *samsara* instead of evil, that is, to speak of a perpetual repetition of birth and death, from the past to the present to the future, by means of six illusory kingdoms: hell, the hungry ghosts, the animals, Asura or the warlike demons, human beings, the gods, and the blessed. Unless one acquires perfect knowledge, that is, enlightenment, one cannot escape from this wheel of transformation, the Wheel of Samsara. Those who are free of this transmigration are considered *Lamas*, that is, the *enlightened* (or buddhas, in Sanskrit).

With regard to gnostic and alchemical symbolism, Jung (1959) deals with gnostic psychology and its connection with alchemical conceptions of quaternity; he recognizes such notions as related to the quaternion of the "philosophers' stone". According to Jung, the quaternity is the ordering scheme *par excellence*, comparable to the reticle of the telescope. It constitutes the system of coordinates that are employed, above all, to divide and order a chaotic multitude of things

FIGURE 16.6 Classic *Tai-Ji* diagram. Wikimedia Commons.

(Jung, 1959). Dealing with antinomies as a form of expressing the content of the conscious and the unconscious, Jung seeks to demonstrate his conception in the form of a quaternity of opposites, as seen in Figures 16.7a and b.

The oppositions between singular and universal, good and bad, spiritual and material (or chthonic) are adequate for the construction of diagrams of contraries. Figure 16.8 is a synthesis of the two quaternities. Individuation for Jung is a process that goes from the personal-horizontal (ego) axis to the transpersonal-vertical axis, that is, to the self. From there derives the choice, somewhat arbitrary, of constituting the polarities of the axes in their quaternities in such a way as to situate related semantic categories such as "unitemporal" and "singular" on different axes (Jung, 1975, p. 214).

Moving beyond Jung's analysis of symbols, in Figure 16.9, we find a binary and dichotomous diagrammatic representation directly related to semiotics, geometry, and algebra (Daghlian, 2009). This figure, known as a *Hasse diagram*, is based on Boolean logic; it can be used in mathematical morphology, a theory which deals with the processing and analysis of images by means of operators based on topological and geometric concepts (Kim, 1997). Peirce also used

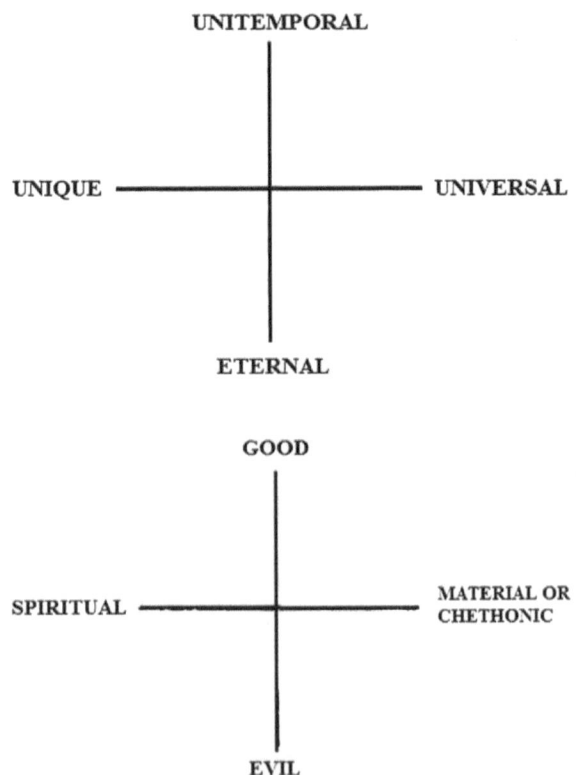

FIGURES 16.7 (a) and (b) Quaternity of opposites; illustration by the author.

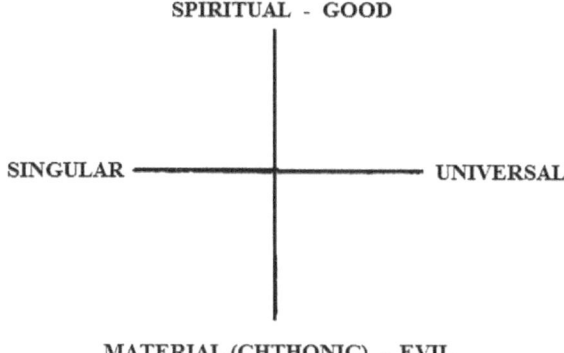

FIGURE 16.8 Synthesis of the Jung's two quaternities; illustration by the author.

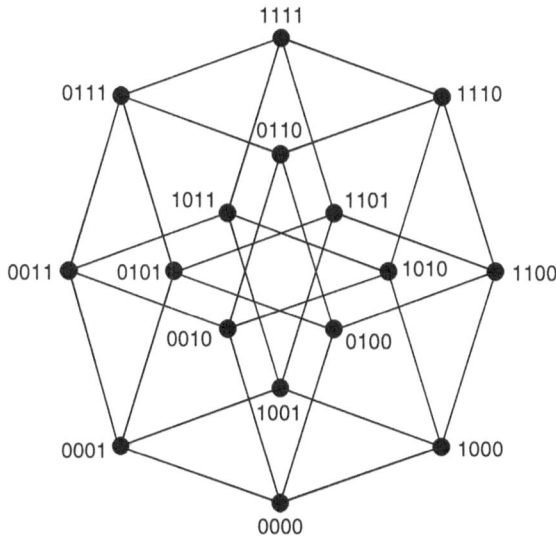

FIGURE 16.9 Haase diagram. Wikimedia Commons.

geometric diagrams in his proposals for binary connectives in the truth tables of symbolic logic (Peirce, 1965, vol. 4, paragraph 268).

The mystic Kabbalah and meta-psychology

"Kabbalah" means *tradition*, and designates a series of speculations commonly considered as part of Jewish philosophy (Mora, 2004). During the Renaissance, Mirandola (1463–1494) sought to give it a Christological meaning (Yates, 1964; Tarnas, 1996). Jewish scholars themselves returned to this manifestation, which flowed in an almost subterranean fashion below the rabbinical hegemony over the study of the Torah. More recently, Scholem (1999) discusses the influence of this mysticism on the prehistory of German Idealism, an influence derived from the writings of Johann Franziscus Budaeus and from the two conceptions of Kabbalah, whether the primordial Gnostic teaching or the dissident strains of this movement.

Beyond philosophy, Jewish mysticism has exercised an attraction on other fields, for example, in mathematics. George Cantor was under its powerful influence (Aczel, 2002). Patai (1994) sought to show the importance of Jewish alchemy from antiquity to the twentieth century, reconsidering the role of the Kabbalah in alchemy and highlighting its importance in a cosmovision in which one finds the constant belief in a "world above" and a "world below" (Patai, 1994, pp. 152–115).

The bibliography on the Kabballah is extensive, but all studies return to the principal source, *Zohar: The Book of Splendor,* an extremely voluminous work of more than 850,000 words (see the selection in Bension, 2006). Jung was quite interested in the Kabbalah and discussed the doctrine of the antithetical sons of God, which influenced Jewish religious speculation and found an expression in the tree of the *Sephiroth* (Jung, 1959). Sigmund Freud knew Jewish mysticism and the Kabbalah very well. Scholem (1999) notes that the granddaughter of an important scholar in the area, Isaac Bernays (1792–1849), became the wife of Freud.

There are, in fact, various indications that Jewish mysticism is constitutive of many aspects of the Freudian vision of the world, but our attention is particularly called to the possible influence of Kabbalistic symbolism present in the "tree of life" (Figure 16.10a). We observe this in the classic diagram representing the Id, the Ego, and the Superego (Figure 16.10b), aspects that recall the Kabbalistic diagram (Gamwell and Solms, 2006). This hypothesis can be made more evident if we consider the original German terms used by Freud: *das ich* (I), *das es* (something in me = this), and *das überich* (what is above and me).

The similarity can be made clear by observing the diagram that represents the spheres in the Hebrew *Sephiroth.* The tree of life is constituted by circles which are interconnected by paths, and each circle and each path represent a divine emanation, or even a form of knowledge (Bension, 2006, p. 289). Although the number of circles is usually ten, some Kabbalistic works suggest the existence of an eleventh, which signifies something like "beyond human knowledge". Interconnecting the circles of the diagram, called the *Sephiroth,* are twenty-two paths, each of them corresponding to a letter of the Hebrew alphabet. These images speak for themselves, and their similarity in terms of an ascending hierarchy is evident, that is, from the most material plane (below) to the most spiritual plane (above). We must point out that in the upper quadrants of the diagram, one finds represented the *Keter* (celestial crown) above the *Binah* (celestial mother of comprehension) and the *Chochmah* (celestial father/wisdom). It is also relevant to point out that the *Yesod* (*Sephira,* foundation) is found vertically below, over the genitals of "primordial Adam", but above the physical world *Malchut* (the kingdom). This hierarchy, in which sexuality is manifested in the *Yesod,* occupies a relatively elevated position in relation to the physical plane.

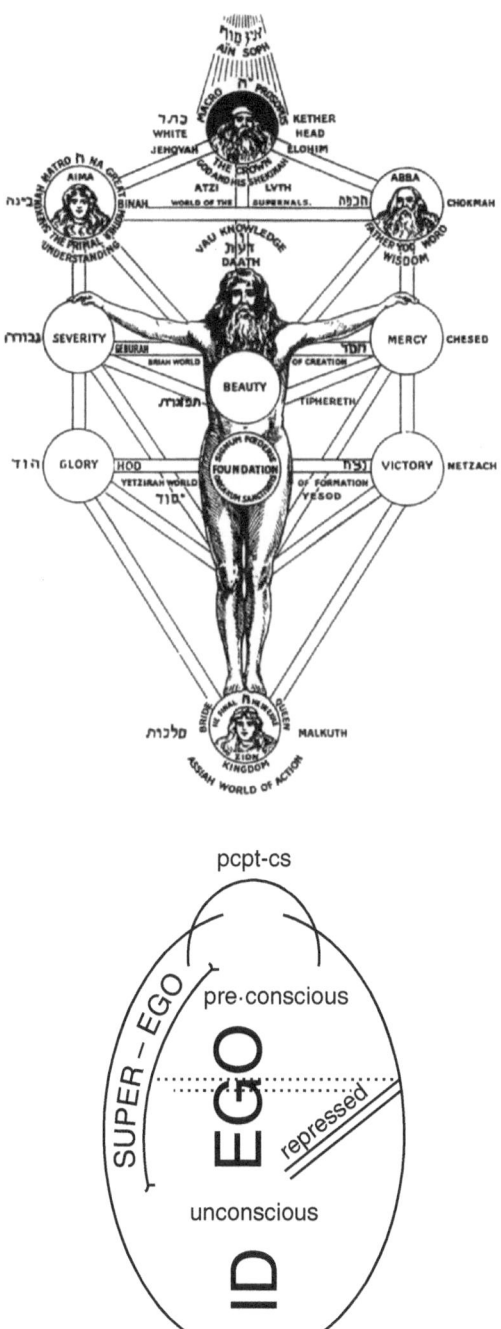

FIGURE 16.10 (a) The tree of life. Wikimedia Commons; (b) Freud's diagram of the personality. Wikimedia Commons.

Final considerations

The diagrammatic character and structural similarity of the figures presented are so evident that they inevitably lead us to conjecture a common origin in the structure of the human mind itself (Ilario, 2007, 2011). For such reasons, it should not surprise us, for example, that an interest in the Kabbalah was shown by scholars of the Renaissance such as Marsilio Ficino, Giordano Bruno, and Giovanni Pico della Mirandola (Yates, 1964, 1966), an interest which is still alive in modern thought in thinkers such as, for example, Henri Atlan (1986, 2005) and George Cantor (Aczel, 2003); nor should it surprise us that the astrological and mystic symbology of the Middle Ages and Renaissance was a factor in the formation of modern science. At least since Empedocles, with his four elements and the eternal struggle between love and discord, there recurs a philosophy of polarity, a kind of perennial philosophy, common to and underlying the great world religions. This fact, ultimately, allows us to see that in mystical disquiet and in the yearning of every cognizant being there is a common ground; in other words, the desire for transcendence has the same nature as that of the desire for knowledge.

Note

1 Enantiodromia is a concept introduced to psychology by Jung, in which the super-abundance of any "force" inevitably produces the opposite of what is expected. It is in a certain way equivalent to the principle of stability in the natural world, in which any extreme comes to be compatible with the idea of equilibrium, in accord with the context in which equilibrium is understood.

References

Aczel, A. D. (2000). *The Mystery of the Aleph: Mathematics, the Kabbalah, and the Search for Infinity*. New York: Four Walls Eight Windows.

Atlan, H. (1986). *A tort et à raison – Intercritique de la science et du mythe*. Paris: Éditions du Seuil.

Atlan, H. (2005). *O livro do conhecimento: as centelhas do acaso* (Vol. II Ateísmo das Escrituras). Porto Alegre: Instituto Piaget.

Bachelard, G. (2002.) *The Formation of the Scientific Mind*. Manchester: Clinamen Press.

Bension, A. (2006). *Passagens selecionadas: o Zohar: o livro do esplendor*. São Paulo: Polar.

Cassirer, E. (1953). *An Essay on Man*. New York: Doubleday.

Coimbra, F. (2004). A arte rupestre do Conselho de Barcelos (Portugal) – subsídios para o seu estudo. *Anuario Brigantino*, 27, pp. 37–70.

Daghlian, J. (2009). *Lógica e álgebra de Boole*. (4 Ed.). São Paulo: Atlas.

Deleuze, G., and Guattari, F. (1996). *What is Philosophy?* New York: Columbia University.

Eliade, M. (1959). *The Sacred and the Profane*. New York: Harcourt, Brace and World.

Gamwell, L., and Solms, M. (2006). *From Neurology to Psychoanalysis: Sigmund Freud's Neurological Drawings and Diagrams of the Mind*. Binghamton and Albany: Binghamton University Art Museum, State University of New York.

Gardner, M. (1958). *Logic Machines and Diagrams*. New York: McGraw-Hill.

Greimas, A. J., and Courtés, J. (2008). *Dicionário de semiótica*. São Paulo: Contexto.

Ilario, E. (2007). *Contribuição para uma gramática especulativa: um novo enfoque em lógica diagramática no campo das ciências cognitivas.* Ciências & Cognição, ano 04, vol 11.

Ilario, E. (2011). *Entre indivíduo-sociedade e natureza-cultura: A constituição do ser – Uma modelagem para a psicologia.* Tese (Doutorado) – PUCCAMP, Centro de Ciências da Vida, Pós-graduação em Psicologia, Campinas, Brazil.

Jung, C. G. (1959). *AION: Researches into the Phenomenology of the Self.* New York: Pantheon.

Jung, C. G. (1968a). *The Archetypes and the Collective Unconscious.* Collected Works, Vol. 3. Princeton, NJ: Princeton University.

Jung, C. G. (1968b). *The Symbolic Life: Miscellaneous Writings.* Collected Works, Vol. 18. Princeton, NJ: Princeton University.

Jung, C. G. (1975). *Structure and Dynamics of the Psyche.* Collected Works, Vol. 8. Princeton, NJ: Princeton University.

Kim, H. Y. (1997). *Construção automática de operadores morfológicos por aprendizagem computacional.* (Doctoral dissertation) Escola Politécnica da Universidade de São Paulo, Departamento de Engenharia Eletrônica, São Paulo.

Leloup, J. (2006). *O ícone – uma escola do olhar.* São Paulo: UNESP.

Lévi-Strauss, C. (1961). *Tristes Tropiques.* New York: Criterion.

Lévi-Strauss, C. (1964). *Totemism.* London: Merlin Press.

Lévi-Strauss, C. (2012). *Tristes Tropiques.* New York: Penguin.

Lima, B. (1997). *Malungo: decodificação da Umbanda.* rev. ed. Rio de Janeiro: Bertrand Brasil.

Mora, J. F. (2004). *Dicionário de filosofia.* São Paulo: Loyola.

Neihardt, J. G. (2008). *Black Elk Speaks.* Albany: State University of New York.

Patai, R. (1994). *The Jewish Alchemists: A History and Source Book.* Princeton, NJ: Princeton University.

Peirce, C. S. (1965). *Collected Papers of Charles Sanders Peirce.* C. Hartshorne and P. Weiss (Eds.). Cambridge, MA: Harvard University.

Scholem, G. (1999). *O nome de Deus, a teoria da linguagem e outros estudos de cabala e mística: judaica II.* São Paulo: Perspectiva.

Silveira, L. F. B. (2014). *Incursões Semióticas.* Campinas: UNICAMP, Centro de Lógica, Epistemologia e História da Ciência.

Tarnas, R. (1996). *The Passion of the Western Mind.* New York: Random House.

Yates, F. (1964). *Giordano Bruno and the Hermetic Tradition.* London: Routledge and Kegan Paul.

Yates, F. (1966). *The Art of Memory.* London: Routledge and Kegan Paul.

INDEX

Note: Page numbers in *italics*, **bold** indicate 'figures' and 'tables', references with 'n' indicate notes section.